NEANDERTHAL EINSTEINS

MJ Politis,

Ph.D., D.V.M., H.B.AR.P.

ISBN 978-1-918002-65-2

9 781918 002652

WOODBRIDGE
PUBLISHERS

276 5th Avenue Suite 704 #944

New York, NY 10001

Copyright © 2020 M. J. Politis

ISBN (978-1-918002-65-2) Paperback

ISBN (978-1-918002-66-9) Hardback

ISBN (978-1-918002-67-6) eBook

Cover Design by Woodbridge Publishers.

Find out more about our upcoming releases and authors at www.woodbridgepublishers.com and sign up for our newsletter to stay updated!

Dedicated to you...

The reader, with appreciation for you being open to this offering, and all those I have known who made this offering possible.

CHAPTER 1

For Grim and Dral, it was another day in the forest, foraging for what the band of other humans in their extended family who had to leave the caves two winters ago needed while living under the sun and stars. And obtaining hides, they learned to put over their heads when there wasn't enough natural brush to keep out the drenching rain and wet snow. The bats and sabre-toothed tigers had reclaimed the inside of the caves. Then, the mountain decided to close up the entrance with an unexpected avalanche. Perhaps such was caused by the imagined creatures who lived above the clouds who, at their whim, provided protection or pain for those below them. But, each species and each mountain had to look after itself. And a species of evolving primates that isn't challenged did, after all, not get any stronger or smarter.

As for getting stronger and smarter, Grim and Dral, twin brothers who had just two winters ago sprouted enough pubic hair on their testicles and contents within them to impregnate the women or girls of the band, were less than mediocre at best. At least relative to the other members of their band of now thirty who spoke the same language and believed in the same creatures above the clouds, sort of in the same way. Most everyone in the band learned very quickly and/or painfully that the easiest way to make enemies of family and friends was to reveal your own real opinion about who those gods really were and who should be their representative chief on earth.

But Grim and Dral weren't able to learn very much else, particularly when it came to being better at anything than anyone else. Rightly or wrongly, they accepted each other's failings at being under-average at hunting game that roamed on the ground, pulling out what lived underwater, skinning or scaling what came into the 'village' (as it was now called), finding stationary eatable wild berries, making implements to cut down wood and pierce living animal flesh, fighting anyone other than their own villagers who came to their cave or camp uninvited, and, of course, finding women to have pleasure with now and who spawned their babies nearly a year later. So much so that Grim and Dral felt like they were members of their own tribe within a tribe, who were kept around, for now anyway, only because they were pitied by their father, Prim, the current chief in most matters.

Prim was a still strong and, most importantly, clever man who had, been born 34 winters ago (60 winters, according to him, when he had eaten too many rotten berries). He blamed Grim and Dral's mother for giving him such defective sons. The other five women who he had pleasured produced strong, strapping, and superior-thinking sons and daughters.

But there were two things that Grim and Dral were good at, other than losing their way in the woods and wandering farther away from home than anyone else had. Or being unable to easily find the camp they were sent away from on a 'special mission' to get food. In such instances, the camp moved its location without its cool and strong hunting boss, Thel, telling Grim and Dral about the relocation, as instructed to do so by the loser lads' father, Prim.

"I think I can hear these berries talking to us," Grim said one sort of fine day to his brother as he reached for a group of three large red balls on a branch hidden by leaves and protected by thorns, feeling very lightheaded and different headed after he sampled one of them.

"Something different than what these roots are saying?" Dral enquired, spotting wild onions that he stumbled upon with his awkward, mismatched feet that, on a good day, decided to work together rather than separately. "But they are saying that... hmmm."

Dral looked at and, as he perceived it anyway, into a cluster of trees down the valley, in the direction where the sun was about to set soon. "These roots are saying..."

"...That if we get bigger berries than everyone else at home, while they get more berries where they went to find them, they will give us more meat to eat?" Grim proposed. "Or not take our hides off us when we're sleeping? Or laugh at us when we take a piss in the woods because they have bigger sticks between their legs than we do? And more hair on their faces and balls?"

"Maybe," Dral countered. "But if we offer these bigger berries to the women, Lolila and Riha. And the others who I don't think are our sisters because they don't look like I remember our mother did. Then maybe they will invite us to share their hides and themselves. We tricked wild dogs into being our friends by offering them food. Why not train women to do the same?"

"That will take more than big berries," Grim related, recalling his own experience with the two women. "Or different kinds of wild onions. But..." Grim was distracted from his next thought by a strange sound in the depths of the valley. It was the song of a bird, perhaps. But it felt like something else. And someone else.

It was a two-legged creature sitting on a moving, four-legged large dog with rabbit-like ears, singing like a bird. 'It' worked its way up the mountain, showing itself to be a bird that walked with a man's head. It carried things on its back in a sac that didn't look like it was animal hides but something that reflected the sunlight. Like a smooth lake with no wind blowing on it.

Grim and Dral both picked up their sticks, hoping the sharp stones they put on the tips would not break off this time. The smaller two-legged creature atop the larger four-legged beast approached, revealing a face that was a man's but lacking any hair on it. He had long, yellow hair, bearing a texture and symmetrical arrangement that neither Grim nor Dral had ever seen. His skin was whiter than any they had ever seen on a human being. There were skins on his body which were not made of animal hide. They were white and bright red, two colors that were only seen in nature when it snowed or when someone or something bled. A strange sounding 'song', which now was being made by some birds and some other creatures Grim and Dral had never heard, came out of the large sac on his back and got louder as he approached. The man added his own voice to it with words that changed loudness in a rhythmic manner, each chain of words having some that sounded like they came from a young boy's throat and some from a big old man's. It seemed to bring pleasure to the brothers' ears and troubled minds, even though they didn't understand the words to the song.

"Are you scared?" Grim asked Dral, clenching desperately to his hunting stick with its naked and perhaps still sharp enough end poised to pierce into the chest of the big dog, its rider, or the sac behind him.

"Yes, but we're not supposed to be, according to our father and leader, Prim, who says he's never scared," Dral replied, picking up his spear with the miraculously still-attached sharpened rock at its end. It wobbled as much as the shaking hand he tried to hide from the singing bird-man. "Our father isn't afraid of anything, right?"

"Except for growing older, weaker, and being told to leave the tribe because he is useless," Grim said. "Something I heard him say when he was half asleep, talking to people who weren't there."

"Dreaming, you mean," Dral said, taking on the top position in their tribe of two, though not abusing the member on the bottom.

"No, it's us who do the dreaming, and during the day," Grim said, repeating the words that had been tossed at and into him as well as his brother so many times by those who could do anything better than the twins could, which was everyone. "Dreaming of a better tribe. A better world. A better father. A better… everything for everyone. Something that…" Grim continued, looking up at the blue sky that was about to turn black soon.

"Dreaming during the day… Something we're not supposed to do because we're lazy and stupid," Dral said through a self-defeating sigh. "And weak," he noted, reluctantly coming back to the world he could see rather than feel.

"And about to be told to leave the tribe if our father gets weaker," Grim pointed out. "Which he is. He shits more than he eats. Though he covers his face with mud, his cheeks are getting paler by the day. And when it comes to lifting anything, he's better at making others do the work than doing it himself. And the shakes he has, which he says is because the creatures above the clouds are talking to him, and only him…"

"Yes, I know… Our father is a dishonest man who told us to never lie, Grim," Dral related.

"Who never taught us how to lie, Dral."

"Something I never wanted to learn, Grim!" Dral asserted.

"Because you were bad at it?" Grim inquired, inviting his brother to come up with an honest answer.

"As bad as you are now," Dral replied, pulling his beloved and fellow accursed sibling in towards the reflective surface of the lake. "Particularly when you say that you aren't scared of that man bird approaching us.

"Who is…" Dral said, noting the bird-man-god stopping. He got off the big dog, leaving it to eat grass, which the large but somehow gentle beast seemed to enjoy. He then walked to a pile of wood on the ground and motioned with a kind smile for the brothers to follow him. "Who is…" Dral repeated, lowering his spear, which lost its sharp tip yet again.

"Inviting us to spend the night with him?" Grim speculated.

The man-bird, who seemed more feminine than most men somehow but in a strong way, placed three piles of food around him. He/she stopped singing and then pulled out what was in the sac that held the strange-sounding birds. Those captured animals were still singing. The melodic sounds of that 'song' came not from winged creatures but from a small strange blue stone with sharp sides on it. He placed it on the center of a slab of wood and something shiny that was elevated from the ground by sticks on all four sides. He then walked to the pile of wood on the ground and set it ablaze with a flick of something in his magic fingers. Then, he set out three piles of food at opposite ends of the elevated platform, placing straight-cut logs in front of them.

"He's maybe inviting us to dinner?" Grim speculated.

"Or as dinner?" Dral suggested, recalling what happened to the animals and patches of the forest when the bright light emitted smoke that got warm when you got close to it and painfully hot when you tried to touch it. He pointed to sharp, flat blades that reflected the setting sun, which the now on-foot Birdman used to cut open two blankets. He laid them on the ground in the manner women did for men back 'home' when they wanted to go to 'sleep' with them, "That man-bird has intense eyes," Dral noted.

"And the big dog with the long tail of hair with the blanket over its back that he's sitting on seems to have kind eyes," Grim noted as the Birdman patted it as a friend rather than something to be killed or skinned. "I say we talk to him."

"Or her?" Dral advanced. "We don't know if this grown man who has even less hair on his face than most of the women in our tribe with the long yellow hair is hiding breasts or balls behind those skins. Or...he/she is maybe both a man and a woman?"

The man-bird got on top of the big dog, kicked him gently with his legs, and proceeded to ride up to Dral and Grim. The brothers each put up their spears, trying to act like the men they were supposed to be. Prepared to kill what and who they didn't understand, like real men. But they couldn't, somehow. The man-bird on the big dog stopped, smiled at them, then uninvited them to his camp down the valley with a wave of his hand, a smile on his lips, and utterance of the word 'welcome' in Grim and Dral's language.

"We… can't…" Dral said. "And shouldn't!"

"We have to get… home," Grim added. "To our people. And you have to get back to your home and your eh…"

"People, yes, I know," the man-bird said, again in Grim and Dral's language. "But destiny calls you both. If you are man enough to answer it. Besides, it is getting late," he said, looking up at the sky. "You can go back to your people when the sun rises a few more times with a lot more than berries and onions. Or if you want things to continue as they are, you can go 'home' now…"

Dral and Grim, for reasons they did not understand, were not so concerned with why this stranger, who looked like no other member of their upright, two-legged species, could speak their language. Of more concern was what was waiting for them at home. A steady decline in the quality of their lives would plummet them into a dark hole. Or the place where those who stop breathing go, which would happen soon enough. Particularly if their father Prim's coughing, shitting, and shaking got any worse.

Grim found himself looked at by the Birdman's big dog. He was normally afraid of dogs, but this one seemed to like him. The dog with the long tail and big rabbit like ears walked up to Grim, seeking a bite of the berries he had collected. After consenting to give the largest berry to the beast, Grim imagined in yet another daydream that it, be it he or she, was inviting him to jump onto his back. He picked up the long, straight, and smooth leather twine that was connected to more twine wrapped around the big dog's head, which was connected to circular rings outside of either side of its mouth.

Meanwhile, Dral's eyes were fixed upon the small ground-connected sun which emitted more light than smoke yet remained in a small place. Grim then found himself hypnotized by the smell of something in his nostrils coming from the food set upon pots over the warm from a distance, hot when you touched it, a sun that didn't move.

"It's called fire," the Birdman said by way of explanation. "Good for cooking food, scaring away unfriendly animals, and other purposes which one or both of you will discover, or spread, very soon. According to this dream, I was assigned to by Life and others to finish so it could become reality," he said. He then looked up to one part of the sky with fear and another with defiance. A crow landed next to him, cawing at his belly. "I will finish what has to be started no matter what you say you want to eat or were instructed to do!" The thus far calm and collected the Birdman grunted with more rage and determination than Grim and Dral had ever seen in one of their own people to the crow with a shaking second finger that merged into a clenched fist.

But the crow, who still seemed to hunger for something in the Birdman's belly, held its ground. The Birdman then yelled the crow away in a language Grim and Dral didn't understand, but felt was very, very intense. It sent the bird away up to a branch above him. The Birdman then let out a fiercer yell, and the crow flew back up to behind a cloud.

"So," the Birdman continued after finally calming down. A warm glow overtook his eyes once again. "Where were we, gentlemen?"

CHAPTER 2

With every answer they got from this Birdman who rode big dogs with big ears and kind eyes, Grim and Dral were challenged with, and scared by more questions. But there was one thing they were sure of. If they stayed close to the fire but didn't touch it, they felt warm. Just like it was summer, even though it was now the season for cool white flakes to come down from the sky rather than warm raindrops, the food they were eating that was cooked by the fire felt hot to the mouth but warm to the stomach. And pleasant to the tongue in ways that raw berries, onions, and meat didn't. And there was plenty of it. Enough to make them feel full in the belly, a sensation which, so it seemed, everyone else in the village back home experienced more than they did, even after a good hunt or better berry gathering.

"Who are you?" Grim asked this man who spoke words in his language that he could understand but not put together.

"And what are you?" Dral put forward as a question that would require a scarier answer.

The Birdman smiled, then laughed. But not in the way the others in Grim and Dral's 'village' did when the two twins accidentally slipped on the ground or were made to fall into the mud after being tripped by someone bigger or more clever than they were. "Someone who is as different from my people as you are and will be from yours," the Birdman finally said after a long delay that scared and fascinated both brothers to more or less equal extents.

"But," Grim said. "Our father said that when we do something stupid or badly, others make fun of us. That's the only thing we do that is of use to anyone else. We make them laugh."

"After you learn what I have to teach you, and teach yourself," the Birdman replied. "You, and your brother, will not be the butt of anyone's jokes anymore but the creators of them."

"But who will the others in the village laugh at then?" Dral inquired.

Again, the Birdman who rode on big dogs who seemed to want to take him where he wanted to go turned his stare toward something behind his eyes, pulling in his lips. Then, after scratching his hairless chin, the outer margins of his lips turned upward. "They will learn to laugh at themselves, or if they aren't brave or smart enough to do that, they will laugh at others who are not you. Hopefully those who can take being laughed at, or deserve to be."

Again, the brothers, who were dumb enough to nearly get lost in the woods while finding berries and brave enough not to go home while they could, scratched their heads. Grim with his right hand, Dral with his left.

"So, this confirms which one of you is right or left-brained, and not a measurement of your skulls," the Birdman noted.

"Huh?" Dral and Grim replied in unison, as they so often did when confounded by life, others, or themselves.

"We work with our strengths first, then integrate them with our weaknesses," their new friend who had provided food to feed their bellies and now something more to feed their souls said with a warm smile. One that Grim and Dral recalled from their mother, who died when they were small kids. Happy kids, as they both recalled, for different reasons. "And you each have different strengths. Different abilities. Different destinies. You, Grim, are right-brained, and Dral, you are left-brained."

"But," Dral said as he pulled his left hand away from that same side of his head. He somehow understood more words the Birdman was saying than his brother did. "We are both alike. When I look into the lake on a clear, windless day, I see my brother."

"And when I look into the lake, I see him," Grim added.

"Which you, Dral, intuited first from your mind, and you, Grim, replied to so… musically with the ability to make the music. Food for the soul," the Birdman replied.

"Music, what is music?" Dral asked with a growl, determined to define the answer.

"Is music an animal in the woods that, if we find it, tastes better than anything else we ate?" Grim speculated. He recalled the 'song' the Birdman and the magic box had both given voice to. He felt the urge to 'sing' with what he had heard and did so. But with words of his own. Then, other expressions that were not words came from his mouth, possibly because of something he ate too quickly, which had been cooked on the fire. "And is music something that doesn't make me belch, shit, or barf afterward?" he speculated from a place inside of him he had never felt and trusted before.

Dral chuckled. He scratched his chin like he saw the Birdman do, thinking that maybe he would be just as smart as him if he acted like him.

The Birdman laughed. "Very clever, and smart, and funny!"

"Why aren't you laughing like the Birdman did?" Grim asked Dral.

"He's thinking about the humor instead of being it," the Birdman replied. "A blessing and a curse."

"What's a blessing?" Dral asked.

"And a curse?" Grim added, with a musical rhythm to his voice that he self-observed.

"Something you both, and I, are required to be," the birdman said, again hiding his eyes and the real meaning of what he was saying. "But first," he continued, pulling out strange tools from the bags made of something softer than leather but harder than fur. "This can help the voice make music."

With that, the Birdman who rode big dogs plucked the twines attached to either end of a wooden plate. He called it a lute. Then he retrieved from his sac a rod with holes in the middle, which he blew in with his mouth. It made different sounds each time he moved his fingers over the holes. He called it a flute. Then he, with reverence, picked up a stack of leather that contained thin pieces of something in the middle that had scratches on it, which he called a book. He looked at the scratches, moved his fingers along them, and said something in a strange language that sounded boring and monotone. Then he said the same thing with a voice that went up and down, loud and soft, and with breaks in the middle of something, all of which made the words somehow sound alive and important. He called it singing, apologizing for being 'off tune' and 'off key'.

All of this was pleasing to Grim's ear. It made him feel sensations going up and down his spine. His feet tapped, and his body swayed to its rhythm. Then, when the 'instruments' were played again, Grim closed his eyes and sang better than the Birdman did.

Dral scratched his chin again, trying to figure out why his brother Grim was moved by this 'music' so much. And why, after he plucked the twines on the lute just like the Birdman did. And was able to blow into the flute, making sounds that he never heard before. He wondered why Grim seemed so moved by this thing called music and he himself wasn't.

Dral felt jealous of his brother, but less dumb than he usually did. While Grim was making more music, the Birdman turned to Dral, picked up a 'book', and told him a story, referring to the marks on thin, large, flat wood shavings as to what the story was. Dral found himself yearning to be able to understand the markings in the book as the mysterious big dog rider did.

"It's called reading," the Birdman said. "Something I can, and am blessed and cursed, teach you. But in your own language, for now. Each mark is a sound, and each collection of marks is a word. And each collection of words is a sentence."

"What is a word?" Dral asked while hearing his lost-in-another daydream-brother make music with the lute, flute, and his own voice. Dral's envy turned into admiration. Particularly when the 'music' Grim was playing and the 'lyrics' he added to the song made him laugh at himself and life. "And what is a sentence?" Dral pressed, impatient for an answer.

"Something you will better understand if I give you this," the Birdman said, after which he hugged Dral as his mother did. Then he pulled away, leaving the caveman turned something else with a strange feeling in the back of his neck.

It felt like a mosquito bite that was under the skin to Dral. A bump that he tried to scratch that itched a bit.

"An implant, which future me says does help," the Birdman explained. "Delivers electrical signals to the brain that makes more connections inside of them. A good thing."

"And if it's a good thing, I want my brother to have one too," Dral insisted.

"Already in the plan, along with how to musically formulate jokes that work," the Birdman replied. "After he's finished with the song which he will sing, you will write it down with other kinds of marks on these leaflets, called pages, so others who will be born after you die, and perhaps reincarnate as, will see how it all started,"

"And so others will know WHO started it?" Dral asked, sensing that the Birdman was as much of an outcast in the village he came from as he and his brother were in theirs. "Others who—"

"—Will keep you and your brother alive, and not banished, as long as you keep them entertained or have a technological skill that they need," the Birdman interjected. He looked at and into Dral in a way that he never felt talked to or at. "A curse and a blessing."

Though still not sure what a curse or blessing was and sensing from the newly activated left side of his head that the Birdman would not give an answer to what either of them was, Dral was determined to find out one thing from this visitor. "What is your name?" he asked Birdman. "And what do your fellow bird men and women call you?"

"Prometheus," the Birdman replied, with a very human voice that sounded, somehow, more powerful than human. And more vulnerable. And more scared of something that would go wrong if he, Dral, and Grim didn't do everything very Right, and fast.

CHAPTER 3

Several settings of the sun later, Grim and Dral finally found their way back home, even though the camp had moved, without their having been told, as part passenger and part rider on more large dogs with rabbit ears and long-haired tails which Prometheus collected from the bush. He called them the horses.

Grim and Dral wanted to introduce Prometheus, their friend and teacher, to the tribe. But the visitor who 'accidentally' found them in the woods when they were lost explained that he had to be somewhere else. Where that somewhere else was, he didn't reveal to either of the brothers, no matter how much they tried to cajol or force it out of him. But he did leave the two lads with sticks to write marks on slabs of thin wood, flat stones, or any other portable surfaces, and a few 'books' which would help them learn how to read and write. And warning them that disaster would happen to them and every tribe if any of these books were thrown into fires.

And with that, Prometheus sent Grim and Dral over the hill to arrive at their wandering tribe's new home, a patch of woods populated by women gathering berries and roots.

Grim strummed the lute with his left hand, his right holding onto ropes, which kept the horse going where the rider wanted him to go, most of the time anyway. He sang a story in Prometheus' language, which he had memorized, being both amazed that he could sing as well as memorize anything. It seemed to please most of the members of the band, particularly the women. And, most importantly, Lolila, the young woman whose legs had less hair than anyone else, and whose breasts wiggled as she walked, then seemed to dance as she worked towards him. This time, Lolila had a smile of welcome on her face to share with Grim instead of a lump of manure in her hand to smear into his face so he would stop staring into her coral-green eyes or gawk at her legs and breasts.

Grim put down his lute, then extended his hand to Lolila, offering her a seat on his horse behind him. She accepted and, with some effort and a lot of fear, was able to find her way to the 'seat' behind Grim. She held onto his torso for dear life when the horse moved as Grim, for the first time, valued rather than lamented his life.

Dral's horse carried him and pulled something behind him called a cart. During the days spent with the Birdman while 'lost' in the woods, Dral had built it his own with some shiny metal implements Prometheus provided for him, along with other parts he made himself from stones, reeds, and wood. It rolled on the ground on something Prometheus called wheels. They were connected by a strong, hard carved stick that Dral had figured could keep them moving together rather than apart, which he called an axel. On top of the cart were boards to make 'portable caves', which Prometheus called 'houses'. Dral saw the designs of such in his head after drawing likenesses of them onto the ground. They could be made from trees that would be turned into planks with sharp stones designed by Dral, which he called axes. Also present in the cart were flints to make fire that would keep the houses warm but not hot enough to burn down. Those fires could cook wild game, plants, and berries, killing whatever was in them when uncooked that made you sick or gave you loose feces. Also present on the cart were three large baskets of plants, which Prometheus helped Grim and Dral gather. Prometheus called them 'medicine', each with a specific use. Dral found himself somehow able to memorize where they were found and what they could do to and for people who were sick.

The men in camp were startled at how well-behaved the horses were, confounded at why they were listening to the commands of the two members of the tribe who were too shy, weak, or dumb to command anyone to do anything. Yet the 'real men' in the tribe seemed scared of the beasts, holding their spears upward at them while retreating backward with each step the horses took.

But they were all most shocked at seeing Dral, one of the shortest men in the village, now three feet higher than they stood, with his chin proudly upward, an arch in his back. Rihi, Lolila's smarter but still not too bad-looking sister, worked her way through the terrified men, offering Dral a piece of raw meat. He accepted it with a smile. Then offered her a stick of meat that had been cooked with his own special spices and, according to Prometheus, something that would ease the pain in her prematurely aching legs and hands, which it did.

But there was one man who was more angered than fascinated or scared. Thel was born with the kind of body that every man wanted. And a man who every woman sought for protection from being cold, lonely, eaten by wild animals, or taken captive by some other band. His muscles were both massive and effective. He stood at least two hands taller than any other man in the village. And when those men tried to put an arch into their back to challenge him face to face, at least one hand higher. Aside from his ability to defeat any two-legged adversary with his muscle, Thel had the kind of eyes that would warn any man who dared challenge or look into them to back down. Those brave or stupid enough to stand up to Thel would suffer the most severe consequences, as he was just as good at slicing and dicing a man or upstart woman with his tongue as with his hand, flint-knife, or, as rumor had it, teeth.

The latter had been used on someone who dared to make a joke during a hunt about Thel's nose being more able to breathe out shit than smell it. That, truth be told, accurate observation cost a fellow band member his life in a hunting accident. His body parts were mixed in with the meat of the mammoth the villagers ate voraciously. Some knew the source of that non-animal meat, some didn't. Two girls had condescendingly rolled their eyebrows at Thel's clever witticisms as a boy wound up with black eyes in another accident. Still, it was 'cool to be cruel' and liked by the coolest hunter, fighter, and fornicator in the village. Who, to be fair, never let the villagers starve or be taken captive. But this time, for the first time in his charmed life, Thel felt both insulted and threatened as Dral and Grim rode their horses into the middle of the campground. Causing everyone in the band to focus their admiration and fear on someone other than Thel, most notably, Thel's two favorite breeding partners.

"What are you doing with these losers?" Thel yelled out to Rihi and Lolila, the two most pleasing to look at and fornicate with young women in the village who, because of such, were, by natural law, his. "These defective pieces of meat who are only around because the old Chief, who is privately ashamed of his sons more than any father could be, hasn't died yet," he grunted, noting that Prim (the lads' father) was nowhere to be seen, and certainly beyond hearing range.

"Grim and Dral aren't defective pieces of meat," Lilila said, struggling to find the words that made Grim seem so lovable and Dral so smart. "They are... hmmm."

"Interesting," Rihi interjected.

"And someone to be proud of... Speaking of which... where's our father?" Grim asked Thel. He gently lowered Lolila off the horse, recalling how when she wanted him away, she pushed him with all of her might, landing him into the dirt.

"And what's that smell in the air?" Dral added, detecting something unusual entering his nostrils. Turning his now, for the first time in public, non-bowed head to the source, he noted a trail of brown stain with a reddish tinge covering portions of the ground.

Grim and Dral dismounted, rushing to the source of the brown and reddish-tinged stream. It led to their father's distantly located place of sleeping, private eating, and (if he was able to still do it at the unprecedented old age of 34) fornicating.

"That's Shit," Thel grunted out. "Big shits. That you little shits should have known about and told us about before you got yourself lost in the woods. Or you shits made deals with other people or ghosts from—"

"—Hmm… shit," Grim interjected, walking around the camp in small circles that got progressively bigger as he poetically waxed on. "What's up with shit anyway? We put green and red things that aren't stinky into our mouths and it comes out brown and stinky at the other end? We eat hard things, and the shit comes out soft, and we eat soft things, and it comes out hard. And when we smell the shit, everyone else's smells stinky, and ours smells sort of ok. But not ok. Maybe if we exchanged noses, we would not mind other people's shit but be sort of ok with our own. What's up with that?"

Some of the men behind Thel chuckled, then quickly put on serious faces when he turned around to look at them. A few of the women laughed a little. Lilila and Rihi laughed a lot, though some of it seemed strained as if it was to please the deliverer of the joke and his serious, really smart for some reason now, brother.

"But you know why I hate shit?" Grim went on, naturally delivering the line just as the group laugh subsided while feeling a buzz under the left side of his skull. "Because it never comes out exactly how you want it to. Sometimes you want an easy flow from the mouth between your ass, and sometimes you want to work for it, and it comes out hard. And if you're a woman watching a guy taking a shit that's real hard, you tell him that it's a whole lot harder and more painful for a woman to push out a baby, who, you hope, doesn't turn out to be a shit."

This time, all of the women laughed from a very natural place. The men scratched their heads. Perhaps thinking about the last hard shit they had or the last shitty kid they had to take responsibility for.

"And ya know what's scary about shit?" Grim went on. "Everybody has to do it. And it's something we can't control. And we all sort of like. And, to be honest, there's nothing more overrated that we bullshit about than an ok fuck, and nothing more under-rated than a great shit."

"No shit," Dral added, with the voice of a wise man far older than his years.

Thel's ears were pounding with laughter from everyone. Even as he perceived it, the wild dogs had decided to train the villagers to feed them and protect them from wilder wolves. Thel clenched his fist in anger and frustration at seeing the two villagers who had been the butt of his jokes for as long as he could remember now become the originators of them.

"But, as we all know," Grim said, seeming to look at every villager, except Thel, as if they were a herd of animals with one mind or at least one agenda. "It's hard to be a shit," he continued, speaking to the band that had become a crowd while looking at and into Thel. But with… understanding, even forgiveness. "Because even though shit floats to the top. It's hard to stay on top and keep the world from turning upside down so that you and everything you built and value do not fall to the bottom. Like shit from your back end that you usually fed to others, and now you have to eat yourselves. And constantly having to look behind your back, which hurts your back. Or having to put eyes behind your head, being careful that the holes you carve into your skull don't make the white stuff inside fall out if you turn around too fast."

This time, even though there was a chuckle here and there, the band members thought, each one for themselves, rather than laughing as a herd following an alpha bull with the biggest horns or a dog with the sharpest teeth. Thel did everything he could to avoid looking into the mirror the short, deformed, big-nosed, and under-muscled Grim had pushed into his face. But instead of reaching for his spear to slay this, to him anyway, the most dangerous beast that had entered the band campground, Thel used a more powerful weapon. "Your father is dying," he said in all seriousness. "And if you stayed here instead of letting yourself get 'lost'—"

"—They wouldn't have been able to do anything," Rihi interjected. "We all go to the place where we don't breathe anymore."

"Or shit anymore," Grim added, trying to make a joke, then realizing that the laugh about to come out of himself and those around him turned to tears of grief. Having realized that he had overstepped a line, he took in a deep breath, hoping that something in the 'magic' he had put out into the air would transform the situation. "But, wherever we go after that time, we don't have to eat anyone else's shit. And if we do, maybe it doesn't taste so bad. And if you ask anyone who has died if there is anything they need, they never say that they—"

"—Your leaving, and staying lost, was responsible for Prim getting sick," Thel, who had in reality not (as Prim expected him to do) told the two brothers the campground would change its location in their absence, blasted out at Grim. "You and your brother have the job to clean up shit in this village. And eat it. And the last time one of us got the shit disease, soon afterward, half of everyone else did. And they died. So, what are you going to do about—"

"—taking responsibility for fixing something we didn't break?" Dral said as he grabbed hold of a pouch inside, deep inside a strange sac made of something that was not leather nor fur. "This medicine can save our father from going to the place where those who don't breathe go. And the father and mother of everyone here went after they got sick. And prevent any one of you from getting sick or sicker."

"So, you can save Prim now and the rest of us later?" Thel demanded, folding his muscular arms.

"I can and WILL!" Dral blasted back, moved to anger yet again by Thel's manipulative stare. "Or eh...I'll...try."

"And if you fail, or things become worse," Thel declared with the authority of the chief he always expected to be. "It will be no laughing matter."

With that, Thel retreated into his own private sleeping, eating, and fornicating tent. He then closed the entrance to that portable cave, feeling the coldness in the air outside. And smelling the shit about to come out of his own ass. It emanated a trace of the odor that was just as bloody and shitty as the trail leading to Prim's place of final resting. But, as Thel was strong, he knew he would survive this visitation from the disease-causing ghosts who moved into the bellies of the living, which would decimate the weak soon enough, leaving the strong alive and in the place they were meant to be.

CHAPTER 4

The battle between the many small creatures who could only be seen by a microscope and the two-legged, progressively less hairy species that had yet to discover that tool was won by the latter after the people, out of fear or desperation or curiosity, took the powder that Dral had offered them. The ground under their feet was once again wet green rather than blood red and fecal brown. Dral had no idea how the herbs Prometheus gave him and instructed him how to find work. Such, the time traveler had related, would take more time and require belief in several medical fairy tales (otherwise called mechanisms of action) till the right one was stumbled upon. The one which, when considered correct, resulted in the healer saving 95% of his or her patients and doing no or little harm to the rest.

The people now listened to what Dral said, even his father Prim, who was still the official ruler of the village, as his shits were well formed and he didn't take shit from anyone else, including Thel. Prim now called his 'accident born' son Runl Dral, Runl being the term for miracle worker of the body. The people followed Runl Dral's suggestions for other ills as well, believing him to be right rather than mistaken about everything one could touch, see, smell, taste, and hear. Belief in the power of the medicines helped make them work. As well as, of course, luck. Or the master of luck, Fate.

Three biological miracles later, the villagers decided to accept Dral's suggestions as a builder. Such resulted in a series of trenches built around the camp to drain everyone's shit, not only what came out of the people who almost died. And those people now lived in tents made of wood rather than leaves and hides, transported from the forest by carts made from more wheels. Seeds from wild plants were put into grooves carved into the ground, with the promise that one day they would produce more food than the wild bushes did, as long as, of course, no one pissed or shit on them. Such gave the wandering people, now stationary villagers, more of what they needed.

As for Grim, he became very good at giving the villagers with (all things considered) healthy bodies what they wanted. But Grim based himself in the worlds he saw in his head behind his eyes far more than those that could be seen by 'real' eyes in the 'real' world, like his brother Dral. Feet made stronger and less painful when walking, jogging and running after game, thanks to Dral, now, with Grim's intervention, danced, both during the hunt and at the village afterward when he sang songs. The most popular subjects of them were things he knew nothing about love, lust, longings, and affection between men and women. The lyrics were taken after things men, women, and two villagers who felt like they were something in between would tell Dral when he was in 'secret' session with them for their medical ills. Naturally, Grim was able to hide their stories of 'pain and glory' in matters of the heart by using fictional names and, interestingly, changing the gender of the person going through such things. They always had happy endings, Grim thinking, of course, that whatever stories he imagined had to come true. Such songs were very popular, as well as jokes he would tell between songs. Grim also attempted to tell stories and write songs about his own struggles before growing into whatever he had become now, and his brother's. To tell the truth in song and music-less joke-story about how things were and how they should was harder to do but more rewarding, for Grim anyway.

Yet such got less laughs and cheers of delight than telling people stories they wanted to hear. The only ones who really seemed to 'get' and like the truth-based jokes and witticisms were his horse, particularly the lament that 'You try to give drowning people a lifeboat, but all they want is for you to sell them more water." And there was that contradiction which seemed to be so true so often, 'act like an asshole and they treat you like a saint'. Such, of course, were observations that Grim would have to work harder at 'packaging', as 'sugar coating around the medicine'.

Prim, of course, approved of Grim's ability to give his subjects what they wanted rather than what they needed between their ears. It made them happier, less able to think about how to change things, and more easy to rule. He gave Grim title as well... Ewit. 'Pleaser of overworried minds'. Grim wished it could be healer of suffering souls, but, in an age when you didn't have a whole lot of choices if you couldn't have the sabre tooth tiger stew for dinner and had to settle for fish, you contended yourself with fish.

All things considered, the tribe needed the transformed Grim and Dral, the twin brothers needed, or perhaps maybe wanted, the tribe to know why they were now so...different. But the two brothers gave their word (the only thing they were really ever good at) to Prometheus to not say anything about him to the villagers, or anyone else. Grim and Dral wouldn't and couldn't reveal the source of their increased intelligence and escalating usefulness to their fellow villagers. Even, and especially, their former most vicious ridiculers and now their pathologically obedient wives, Rihi and Lolila.

"The purpose of making these two more kind heart than clever cavemen into geniuses was so they would breed and enhance the genetic pool," the time traveller's currently four-legged companion said to Prometheus from behind heavy bushes atop a cliff that no one dared to climb. "It's only a matter of time till they, as future generations would say, spill the beans to Rihi and Lilola or their children," she continued, pointing to Rihi and Lilola, now the best-clothed women in the village, being served food by the campfire by both men and women who were clad in patches of furs rather than full-length ones. "Or the children of all of the women in the village who they will make pregnant and haven't yet."

"Grim and Dral are loyal husbands, Athena," Prometheus replied in defense of the two candidates who had sojourned into the outskirts of the forest, 'the people' had previously never left. "Kind and caring men."

"Who are now owned and manipulated by selfish, jealous bitches in training," replied the goddess, currently taking form as a female squirrel so as not to be detected by her bosses above the cliff or the forefathers and mothers of her future worshippers on the ground. "We should have chosen two men of higher intelligence for this Mission."

"There is intelligence of strength and heart," Prometheus said. "These two men, Grim and Dral, who left the forest that protected them—"

"—because they were lost!" Athena chirped back in squirrel talk, her eyes enlarged, her tail upward, her teeth about ready to take a real bite out of Prometheus' overactive tongue, shaking cheeks, and determined eyes. "They have NO idea of what they are getting into!"

"Something we all share when we do something Right, and needed, that makes us… uncomfortable and less secure," Prometheus replied. "Which means… hmm." The extraterrestrial humanoid from another time and place who would be, by accident or intent, called a god in future times, put his very human hand over his mouth, scratching his hairless chin.

"'Which means hmm' what!" Athena demanded, lowering her tail. And edging her way in towards Prometheus. "What other plans did you set in motion without asking our boss and his boss about the species we are assigned to take care of?"

"So that they can become better than we ever were, or maybe are?" Prometheus said. "Which meant that… for them to be saving US from ourselves one day, they, Grim and Dral anyway, have to become less uncomfortable and less secure."

"And… challenged," Athena replied. "Challenge being the thing that expanding and creative souls need most of all."

"True enough."

"But not for me," Athena chirped back in a voice more like a lioness than furry-tailed rodent that could be easily eaten by the former. She leaped up to a higher branch, curling up, helping herself to nuts and berries that were not spiked with anything other than mind-non-altering carbohydrates, proteins, and fats. "I am still a goddess And don't want any more challenges. Neither do the other collaborators we came to this wretched planet with who, hmmm, are just' "

"Hmm, just what?" Prometheus asked.

"Who just left the project," Athena chirped out. "But of course will join it when and IF it succeeds," she smiled on the way out as she ran up the tree, then, with a flip of her material squirrel wings, bolted up into the sky, after which she jet-propelled into and as a beam of light that disappeared behind the clouds, this time not leaving behind a piece of material plane shit on her former lover, and current explorer partner's, head.

CHAPTER 5

The collection of cave dwellers who had been forced to live in the open small, then bigger village jad... problems that came with progress. Newcomers from adjacent regions noticed that they were prospering and doing so under the sunlight. Some came willingly, offering meat, furs, different kinds of spears, or a breedable daughter as a gift to the 'settlement'. Prim of course, got the lion's share of those gifts, along with a bowed head during the giving of such.

Some of the newcomers were 'encouraged' to join as second-class 'citizens' after Thel, sometimes with and sometimes without Prim's authority, and more often than not without his knowledge, went out on raids, which involved killing the settlement's most dangerous enemies. He captured those who chose to become servants rather than food for the vultures or stew for the conquerors. It was a fair exchange of security, comfort, and food to avoid the terrorizing phenomenon some called freedom.

Tales and legends about Prim, of course, grew, many put into song by his son, Grim. But mostly by Grim's students, who wanted to rise above the social status of their teacher. Indeed, tales of Prim remaining a strong, smart, and, all things considered, not unattractive to look at old man of the unprecedented age of 34, then 44, then 54 grew faster than the seeds planted in the ground that bore edible fruit and vegetables for the settlement, as well as the inedible weeds around them. Sickness seldom led to death. The most common cause of death now was by the hand of humans against other humans rather than Mother Nature culling out or challenging the species with the opposable thumb that dared to change rather than work with her. According to fact and fable, there was no place more prosperous and happy than the settlement and no ruler who triumphed more over death by any cause than Prim. Until a lone hunter from another valley, far beyond the current borders of the settlement and its hunting grounds, walked proudly into the settlement just as it was about to officially call itself 'Primton' at a celebration feast that everyone attended with varying degrees of willingness to do so.

The lone hunter was traveling alone. An independent soul who saw no need to conquer someone else to avoid being conquered. It was most alarming to every male 'citizen' in Primton and fascinating to every un-owned female, a woman. She carted with her a single spear with a shiny sharp blade accompanied by an artistic multi-imaged down design on its shaft in her right hand. A slain animal the size of a deer lay over her shoulder. The coverings on her feet were hardwood rather than soft leather.

"An interesting village you have here," commented the physically fit but not overly muscled 22-year-old female humanoid with long, untangled brown hair and a clean, scarless face while being stared down by thirty spear, club, and now knife-holding men twice her size. She strolled into the settlement like she owned it, laying down her spear and game, sure that no one would take either of them. "And interesting people in it," she continued as she smiled at still small-framed, unarmed, and comfortably clad Dral and Grim.

Just as the edges of Grim and Dral's lips turned upward nearly to the level of their wide-open eyes, they were brought back into a serious, expressionless 'profound' position by Rihi and Lolila. The former reminded Dral of his honored position as the settlement's healer by putting her arm around his elbow. The latter discretely stepped on Grim's foot before it was about to tap out a song he no doubt was composing about this strange, strong, and alluring female huntress, who combined the best elements of being male and female.

"And you have very interesting animals," the lone huntress said as she approached the horses Grim and Dral rode into the settlement when it was merely a mobile campground. She blew into their noses, petted them, and then noted an enlarged abdomen on the one who didn't have a penis. "Two animals who are smarter than we ever are, with one more on the way," she noted.

"I… eh… knew that they were pregnant!" Dral blurted out. "I was going to save it for a surprise for all of you!" he announced to his congregation. "Once, I was sure that there were no complications with the birth."

"And for that birth, and for the mother so she stays healthy, I composed a song, which, maybe on this great day in this good village which has become a great settlement, I can sing now," Grim interjected as quickly as he could. But before he could improvise the first lyric from his every creative mouth and pluck the first multi-stringed sound from the lute, which his fingers somehow knew how to put into 'chords', the Huntress spoke up.

"A great song, I am sure, but our songs are better, have more notes, more chords, more thought-provoking and emotionally-moving lyrics, and serve as well as please the listener, even those who don't understand the music or the lyrics," the Huntress interjected. She wandered around the camp with more bold confidence than any man and more musical grace than any woman. "Our citizens are freer," she said of the slaves and their new owners. "Our wagons are faster and more durable," she commented regarding the three carts that were functional and two in need of repair. "Our houses are bigger, and warmer," she went on, looking with pity on the dwellings which the 'Primians', as they now called themselves, thought were the best living places imaginable or buildable under the sun. "Our bodies are healthier," she continued, taking note of the sick who Dral was able to save from death but not coughs, limps, pains in their hands, or limitations in the ability to see and hear the world as it is. "Our elders are… older in both legend and fact," she proclaimed with complete ease, yet with intensity of voice and thought as she gazed at and into Prim. "And..." she concluded, turning around to the congregation after walking through them, eying with a condescending grin the weaponry Thel and his men were poised to insert into her. "Our spears, clubs, and what you will soon call small swords are not only better than yours but are used to kill animals, not people."

Thel grunted, growled, then grabbed hold of the upstart Huntress, putting a choke hold around her neck. "You say untrue things! You are a witch! Who brings bad magic into this place! Who I will kill!"

"Maybe, or maybe not," she said, without a trace of fear in her eyes or a drop of sweat on her brow. She proceeded to grab hold of Thel by his enlarged, and still (because of Dral and Grim's interest in their work rather than women, and their wives not wanting above all for other women to be interested in them as people) overused testicles.

After a moment of shock, Thel found himself flipped around, hitting the ground with his back then head. "I can't move my feet!" he whimpered like one of the helpless slaves he had conquered.

"The first fall you have in life is always the hardest, but you'll live," the Huntress said. "As Rhul Dral can attest to."

Dral rushed over, pin pricked Thel's feet, then arms, then gave him a potion that made him feel better, or breathe easier anyway. "You will walk again," he pledged to Thel.

"And dance again," Grim added.

"But not on anyone else's back while they are alive or their belly while they are dead," the visiting woman said to Thel. "You promise me that, and I will see that the magic these two men you still hate, fear, and do not even try to understand will make you better." She turned to Dral and Grim, winking at both of them at once, making each of them feel like the connection was for both of them, collectively and individually. "Right?" she said.

"Of course," Dral said as he reached into his bag of 'magic tricks', pulling out an elixir. He stuck it under Thel's tongue.

"And for the medicine delivered into the ears rather than the mouth," Grim added, strumming a new tune on the lute. Then singing lyrics about healing of the soul in a language he invented in his head that he would translate later, if at all.

"Indeed, yes," Prim added as if emerging from nowhere.

Thel emerged from the ground with a painful leg, aching head, then arms that embraced Grim and Dral's two boy-sized bodies with one man-sized hug. Thankful that he could feel his limbs and that they were working.

"My sons carry powerful magic," Prim announced proudly. Indeed, he was as proud of his boys as he was glad to finally see super-strength Thel, his rival, who had never gotten so much as a single cut in battle with man or beast, brought down to being 'mortal'. "Yes, my sons carry powerful magic in their hands and heads."

"But..." the Huntress pointed out, after which she purposely delayed the rest of her claim. "They do not have as much magic as we do in our settlement." With that, the woman from the 'better and more accomplished' place she declined to name or locate picked up her admired but not touched spear and uneaten game. Then left the village, disappearing into the woods on her shoes with wooden soles. She was followed by no one and nothing except the curious eyes of the Grim and Dral. They were both determined to follow those footprints wherever they led, without, of course, telling anyone else.

Prometheus admittedly was new to shape-shifting. His penis was tiny but still intact in front of the vaginal opening he had made for himself as the Huntress, which, perhaps if he was better at this skill, could have led to a uterus capable of bearing children. But, experiencing life as a female humanoid was instructive to him, scientifically and artistically. Indeed, he considered the logical fact and expansive methodology that if 'healer of the body', Rhul Dral would spend some time living as Rhul Dral, even in clothing and social status, he could experience what women go through biologically and otherwise, and therefore be a better healer of their bodies.

As for Ewit, pleaser-and-wanting-to-be-server of the soul, Grim, that was another story. Grim, being more imaginative and artistic than the now technologically obsessed and, to a limited extent, super-skilled brother, seemed to enjoy wearing female attire when acting out female characters in song and non-musical theatrical presentations. He kept them on longer than the performance required and often maintained that attire when in his private music-writing room. A good thing as long as he didn't become so 'different' than the men in the still 'men must be men and women must be women' that he would get raped up the ass by Thel et al. Or, worse, be scolded for not investigating and USING the female side of his Soul by his traditional and psychologically powerful, 'every man's man' father, Prim.

But, such experiments would have to wait. Prometheus, while delaying his transformation from female Huntress into male explorer, noted lights in the sky from flying craft that were not from his planet but the other 'advanced' civilization, which was on a collision course with eventually destroying itself in a war those two worlds were engaged in. But even more importantly, other lessons had to be learned by evolving mortals and 'gods to be', against a time clock that Prometheus had set into motion. Such wasn't in accord, of course, with the timetables of his boss and father, Zues, Mother Nature or Spirit, big S, within every being which ruled all gods, no matter what planet they came from. Spirit which, perhaps, would be discovered, and not misrepresented, by humans on this third planet from the sun.

But Prometheus knew firsthand that a mind that thinks it is smarter or wiser than anyone else's not only halts its own growth but sprouts seeds of ignorance, which inevitably germinates into cruelty. He knew that a healthy dose of feeling second-rate or inferior makes one not only better than what circumstances require but enables someone to be better than who and what he or she is. Assuming, of course, that such humility-infused intelligence evolved into effective compassion before it was subverted into depression and learned helplessness at the hands of... less committed or less spiritually and mentally evolved humans. Such 'lower' two-legged creatures who, due to circumstances beyond Prometheus' or even Athena's control, were still in overwhelming numbers. And they had at their disposal multiple tools to use against those seeking to be, needing to be, or forced to be... enlightened.

CHAPTER 6

It had just been another morning for Dral, giving the native Primtonians in the settlement and, when he could, the newly arrived or captured people what they needed medically. Each case now seemed harder, even though he was a better healer. Supplies of medications Prometheus had left him were dwindling, and the fields that were supposed to supply more raw materials for such were not as plentiful as before. The books left to him describing how the human body worked didn't go into as much detail as to what happens when things go wrong, requiring him to use more logic and intuition than normal.

More importantly, Dral was still not yet able to convert sick people into healthy ones, or sometimes even dying people into merely sick ones, to the extent that healers in the Huntress's Settlement could. But he was the best doctor in Primton.

After medical duties were finished, he had set upon the tasks of repairing the houses and tools used inside of them. Such came with mounting challenges, including how to bring fire, and perhaps even water, safely into those dwellings in preparation for the upcoming winter. And there was the issue of devising better carts to bring in materials from the woods and cliffs around them. Still, every time he worked within or beat the laws of physics and nature, he felt that his accomplishments were second-rate relative to the Huntress' home.

As for Grim, he dived deeper and deeper into himself to come up with stories, jokes, and songs that would make people laugh and think, seeing the importance of both. And, of course, passing on what he did or didn't know to his students. Particularly the sons and daughters and wives of the most well-off Primtonians who he was required to convert into musicians and singers. Even though for most of them, converting their blisterless fingers and obnoxious voice-boxes into conduits for music likable by anyone except themselves was as hard as turning the manes on the horses into wings that would enable steed and rider to fly into the clouds, as was probably the case with people and horses in the Huntress' 'kingdom', or perhaps 'queendom'. Still, for the moment, he was able to give the people who came to him laughing, smiling, and singing enough to not be thrown off a cliff as they chanted his death song.

When Grim and Dral were finally allowed their mid-day lunch breaks, it was just another late afternoon with enough light for the two overachievers who had not thought of themselves as underachievers to pursue higher and more personal agendas.

As they had done more times than they remembered, Grim and Dral followed the footsteps of the Huntress into the woods. They were easily followed, of course, as her footwear was something different than anything left by man, woman, or beast. And on occasion, the genius lads found something else left along the side of the path that the Huntress left as gifts for them or accidental droppings. Sometimes, it was a parchment with mechanical drawings of something buildable by human bodies. Sometimes, she 'accidentally dropped' drawings of what perhaps lay under the skin that made those bodies able to do the building. And sometimes the Huntress left behind musical instruments which Grim somehow was able to fix, sometimes with and sometimes without his brother's help. All led, many times, to a circle of rocks in a holler hidden by a thick bush, where the tracks of the elusive Huntress disappeared. "So, should we dig deeper to find out where she went to?" Dral asked Grim as they both gazed into the black hole they had dug so deep that they could not see the bottom, even when the sun shone directly into it.

Common dirt came out of the hole each time they dug, but Dral's now surgically instinctive fingers could sense some kind of 'buzz' to something inside the circle of common stones. Grim's feet and gut, which could feel vibrations of music played by anyone, even himself, sensed some kind of 'throb' to it. Both brothers felt those sensations more than normal now. "Perhaps this has something to do with a place that's always doing things better than ours does. And we do."

"Or can, unless we find this kingdom. Which..." Grim said as he turned around. He then strolled on and 'listened to' the ground around the circle of rocks with his bare feet. He used his imagination to allow him to feel some kind of 'music' or 'drumbeat' under the ground. This time it seemed to lead him to the direction where the sun rose rather than set this time. "Maybe she went in this direction, with different feet," he speculated. "Or a really big jump that some would call flying, maybe propelled by a really big fart out of her beautiful ass," he mused.

"Which theoretically could enable her to fly, IF she was able to have wings that moved in a downward, back, then upward movement like that of a bird," Dral speculated with expressionless clarity, ignoring or, as Grim perceived, unable to pick up on the fact that the fart-flying remark was intended as a joke. Something that Grim was becoming better at making over time, yet he was unable to laugh at any of his own jokes. Or, for that matter, anyone else's. And as for the stories Grim devised and music he composed, he was equally incapable of feeling emotion from them, other than the 'sense' that the song, story, or joke 'worked'. As the eye of the tornados of imagination he created, indeed, Grim was unable to feel or know what it actually did. Or were. Meanwhile, Dral, who had become so good at things technical, lost all connection with things artistic.

But there was another reason why Grim and Dral were so intent on finding the technological and artistic Paradise that the Huntress had to have come from. "You know, she was smiling at me but only looking at you," Dral said. "There is some kind of connection between us which could advance our ability to combat the cruelest intentions of Nature to levels that—"

"—You are wrong, brother," Grim interjected. "She likes you, but she loves me. Someone who knows what love is and could be."

"Love," Dral replied, scratching his chin in the same manner that Prometheus did upon their 'accidental' meeting with him. He remembered what he once felt, or still felt, for his miraculously-obtained dream wife, Rihi. "A psychological phenomenon which requires better definition involving expectations between both parties and methods of delivery of such," his summary of the situation.

"Love is something you feel and can't understand," Grim pointed out with open palms and an open yet still hurting heart, recalling his own oscillating internal feelings for Lilila. "But, would you rather have a woman who understands your mind or connects to your heart?" he advanced, hoping that the real meaning of that inquiry would connect to Dral's still hopefully Alive, big A soul. A soul that could not lose himself in any fictional story or transformed by any song, at least that Grim could compose.

"A woman who understands your mind and connects to your heart?" Dral inquired.

"And who you are yourself around, your Real self," Grim added.

"Who the Huntress probably is, for me," Dral concluded.

"And me?" Grim proposed. "If a man can love more than one woman, maybe a woman can love more than one man?"

"As long as that woman who… looks familiar," Dral said as he proceeded to walk to the East. "Maybe feels for us?"

"Like all women do," Grim said, trying to convert that very real remark into a joke but being 'stuck' at 'profound' and accurate, following his brother.

"Yes, as long as that woman can love only two men and not three," Dral added with a smile. Once, he was again working with the third brain, and now Mind, that the brothers always shared and needed to rediscover. "That woman who—"

"—That woman who what?" the brothers heard from behind them in a familiar female voice.

"Who these idiot geniuses think they can measure up to," a second woman with the first said.

Rihi and Lolila approached their husbands, anger in the former's face, rage in the latter's.

"We were... just, ya know," Dral blurted out of the side of his mouth with quivering lips.

"Out looking for better medicine, technology, and..." Grim added. "And fresh jokes from the squirrels, new songs from the birds, and—"

"—Specimens to breed with?" Rihi interjected.

"While you BOTH have responsibilities to your own breeding partners first?" Lolila added.

In unison, the two women opened up the furs covering their enlarged bellies, patting them.

"Who will be arriving, when?" Grim asked, unable to differentiate between an overfed belly in a woman and one that was soon to sprout yet another Primtonian.

"Soon enough, for you," Dral said to his brother, assessing Lolila's exposed body, which, apparently, Grim had not seen nor felt in a long time. "And," he went on with a forced smile as he tried to convert the fear in Rihi's somehow 'lighted up' face into assurance. "A bit later for me."

With that, Grim and Dral escorted their wives back to Primton. The former wondered who the real father was. And the latter hoped he would never find out. Dral began to prepare himself to see if his son or daughter would look like him or someone else. Something that his father Prim, perhaps, had experienced as well several moons before he and his twin brother had crawled, or was pushed, out of the womb.

CHAPTER 7

More time passed. Primton grew from a settlement to what they called a City. Very soon, there were more builders, farmers, and hunters than warriors, as the latter were not needed as much. Prim remained the leader, but the people discovered, with Grim and Dral's help, a new form of government. But Prim's sons did what they could to maintain their father's physical strength and ability to use charm to get others to do his bidding when he couldn't do it himself. The old man of 34 years was approaching the time when he would go to the beyond world no one in this one had ever visited or at least come back from. His hair was becoming white then started falling out of his head. The ability to remember things fell out of his head as well. But, Prim did allow the new form of government outlined by Dral and made popular in song by Grim to establish itself.

They called it democracy. Rule of the people. Most of the time, it served rather than subverted the people. As long as Grim and Dral were able, in ways that were not recognized, to serve rather than please them. And after giving them ten things they wanted, giving them at least one thing they needed.

As for Grim and Dral, they kept looking in vain for the Huntress' ideal civilization as the borders of Primton expanded. The brothers, whose skills were sometimes complementary and sometimes shared, were still convinced that, from what they saw of its 'garbage' left behind in the woods and legends from other hunters who passed through Primton, they had to become competitive with the Huntress' homeland. Their new goal was to be worthy of being with her, and this required renewed efforts to get their shit together at an escalating rate. If they did good today, they had to do great tomorrow.

Such was an obsession that made them effective as advanced citizens of Primton but strangers to the people they served, and those they wanted to love, most particularly their children. Who were told by their wives that their hard-working fathers, who built great homes for them but were seldom in them as fathers, cared more about everyone else's family than their own.

Grim and Dral had given up trying to convince their wives that selective compassion for 'family', or one's own, at the expense of everyone else's led to wars that destroyed everyone and everything. However, the brothers did make some progress. They hoped and prayed (to Spirit they felt more connected to than the people it made), teaching their sons and daughters the value and necessity of universal compassion. Caring for all equally, each giving according to their ability and taking according to their needs.

Such had been observed and noted by Prometheus and Athena who, on an unusually cold and damp winter day, did not shapeshift into anything else as they continued to do maintenance on the circle of 'common' rocks which for them was a porthole. An entrance to another time and dimension that, thankfully, was not discovered by anyone other than Grim and Dral.

"Those books, drawings, musical instruments, and other tools that these 'accidentally' less hairy and arrogant primates found along the trail to this place," Athena said as she moved the North stone into the correct position so as to coordinate with the vibrations emitted by the ones on the East, West, and South aspects of the circle. She used very human muscle this time, as she had used up all of her telekinetic powers for the month elsewhere. Her body smelled of sweat and felt dry. Her long blonde hair was in tangles, impregnated with dirt. Her face showed wrinkles which her make-up was unable to hide this time. "It makes evolution too easy for them," she said through a phlegm-filled cough that emitted a tinge of blood within it more so than normal. "Remember what happened when you, or rather, in our naïve optimism, 'we' just gave them away?"

"Yes," Prometheus said, his body in better shape than Athena's on the outside, but feeling a weak aching in his legs, congestion in his chest. He retrieved from his satchel a manuscript with his prematurely or perhaps merely weather-caused arthritic hand, looking at it with his still focused bloodshot eyes. This one was in the language of words, which Dral was versant in. It also contained the language of mathematics, in which Grim had become an expert in thinking. It described some of the physical laws of matter as it could be touched and felt and the inner workings of small particles that made up matter, which would enable one to transform matter into energy and vice versa. And enable one to be independent of the restrains of linear time as well as the discomforts of physical things in their own time. "These two misfits, and perhaps their students, are improving on what we left them. They are using what we taught them to teach themselves to be..."

"....better than we are, or are becoming?" Athena shot back.

Prometheus scratched his chin, then after a reflective pause, replied with a wry smile. "Yes, we can only hope so."

"They will ignore us and destroy us if they get too smart too fast!" Athena said.

"Or save us if we help them become wise and self-sufficient. And... yes, I know, use the tools we gave them to build rather than destroy."

"But to build what?"

"A world better than the planet we came from, and are in the process of destroying ourselves, without the help of any other species," Prometheus recalled how mass and energy inter-conversion was something that had enabled small minds to become destructive ones. He looked at the sun, recollecting how smaller fires on his home planet had converted so much of it to ashes. "Perhaps if we can show these advanced primates how to change their world, they can teach us how and why we should change ours. Or—"

"—Share their world with us," Athena replied with her alluring yet thought-executing trademark eye-roll. "Is that a dream or yours, or something that Grim dreamed up in the imaginary world between his ears?"

"He doesn't know about who are what we are," Prometheus said.

"Not yet," Athena shot back. "And when they do find out our most important secrets, you know what we are required to do to, and with them. According to the laws of nature, neither Zeus nor that Spirit, which he and you seem to be looking to do the Will of, can change. Even Grim and Dral can't change the laws of Heaven and Earth. And what if those two, or the people they are ruling now, find out that the 'rush of discovery' they feel inside their tiny Neanderthal heads are caused by electrical currents making their brains sprout new connections. And they figure out how to make more powerful brain implants than the ones YOU put into those two still naïve idiots!"

"Yes," Prometheus replied, looking downward with regret. After which, he looked to the sky. He felt himself becoming, yet again, a channel for something bigger than himself as his mouth let out. "But, heaven watches, earthworks. And once a wave of energy is created, it is never destroyed." As for what that energy was, Prometheus dared not share that with anyone, even Athena. She had as much to lose as he did at the hands of their bosses and the ultimate Boss of them all. As for entrusting Grim and Dral with the future of humanoids on this most ineffective planet, that would have to wait. The experiment had turned into a gamble now. Where the fate of many worlds was at stake. Grim and Dral were the most expendable chips on the gaming table. In a game Prometheus was responsible for setting in motion. It was an awesome burden for someone who sought to honor two agendas. Above all, do no harm. And make as big a positive difference in the universe as you can while you can. Both HAD to work together, now more than ever.

CHAPTER 8

It had to happen sometime. The oldest man in Primton, or
any other campground village of hunter-gatherers in the
known world about to be a 'town', eventually was now
gone, lying on a hard wooden plank made softer by a
covering of furs and straw. A smile emerged on his
weather-beaten, wrinkle-covered, and battle-scared face. At
the old age of 42, Prim finally became one of the humans
who seized breathing air. No one was in his house hut to
hear what came out of his mouth with the last breath. But
something significant did come out of his ass. "The hardest
shit I ever grabbed hold of," Thel said regarding one of his
most recent jobs of 'maintenance medicine' dispenser for
the elderly and feces remover for everyone else. "But when
I put the detritus in the fire or holes to ferment into
fertilizer, it gave out the strangest fragrance into my nostrils
that have experienced in many plethoras of breaths," he
noted as the first person who found Prim dead. He was

followed by two others who voluntarily came in to check on him a hundred breaths of the living later.

"Interesting that he speaks like us or is trying to," Grim whispered to his brother regarding the upgrade of verbiage from Thel's customary one-syllable vocabulary, out of ear range.

"But Thel is not like us," Dral added.

"I was talking about Prim, our father," Grim reminded his brother. "Who—"

"—Will have to be replaced," Thel reminded Grim and Dral. "And as his two most popular sons, you both—"

"—Don't want to be part of any club that would have me as a member," the art-loving and politics-hating Grim self observed himself coming up with on the spot.

"Me either," Dral added, not quite understanding the illogic of that statement. But he knew that his brother's non-linear mind somehow got from A to B in a circle that was very often faster than a straight line. And, of course, Dral knew, very logically, that to let the 'common citizens' know that there was discourse between the most 'developed' pair of souls in Primton would result in death for both of them.

Indeed, Dral and Grim found themselves unable and unwilling to share most of their thoughts with any of the other villagers. And not only because the two brothers were faster and, they hoped anyway, deeper thinkers than anyone else. Having learned to proactively design a life where there was minimal conflict, the twin geniuses were able to convince their wives and children that they were loved.

But the truth of the matter was that Dral and Grim tolerated, served, and pitied the family members; biology required them to be most responsible, as well as the villagers that the moral code the two brothers had adopted and created were self-assigned to look after. There were no two-legged upright beings looking after the genius twins, most particularly now that Prim had passed away.

Prim had now passed into a dimension that Grim was forced to devise believable stories about to comfort his mourners. Dral found himself not believing in any of the tales Grim came up with, as there was no proof for any of Grim, or anyone else's, description of the afterlife.

They would miss, and now need, Prim to say something to make Grim and Dral feel ok about themselves when they did 'good', and good about themselves when they did 'great'. But with no Prim around, the only voices in their head were the 'you're a piece of inferior shit' rants, which were silenced only by having done something miraculous, for a day or part of a day anyway.

There were many things that Dral, a genius inventor-doctor who made biological life possible and comfortable, and Grim, 'magical' entertainer who gave people many tailor-made reasons to live, didn't share with their fellow Primtonians, and the plethora of villagers who flocked to the campground they converted into a settlement, then a village-city in the hope of becoming Primtonians.

When the sun went down, they knew it would come up the next day. And when the days got shorter as the weather got colder, they knew the day when the sun would begin to shine longer each day. When patients felt pain, they knew that it would subside, eventually, and be replaced by pleasure. They could measure how long it took between a man and woman 'sharing the same blanket' at night and the day when said female would pop out a newborn native-born Primtonian. They knew that disputes between husbands and wives escalated between love and hate and that indifference was the sign that the relationship had slipped into toxicity or its natural ending. They knew that Nature, even when 'she' was most vicious, never gave you a problem without a solution. And that Nature was not really a 'she' but an impersonal congregation of various biological experiments competing for superiority, with, for now, mankind on top, assuming that greed and fear would not overcome the species.

But they didn't know how to experience real human emotions, especially the happy ones. Dral began to feel 'just ok' when he miraculously rescued a patient from death or debilitating disease, having been so good at preserving life that he felt dead himself inside. Grim became a master at making people laugh, even at the saddest of times, but was unable to share a laugh with anyone. He had come up with so many stories in words and songs that the wonderment required to write new ones was gone. The two brothers' smiles amongst humans were now forced, their hugs mechanical. They bypassed happy at each opportunity out of a sense of duty, habit, or perhaps shame for letting their father get so sick.

Yet, they discovered Bliss, experienced after each discovery in their very different areas of exploration and service. They imbibed a meal of satisfaction alone. They heard the sound of Silence as loud as the thunder. And they had to be satisfied with that Inner Applause for periods that lasted less than the average rainfall.

So, what do we do now?" Thel asked Grim and Dral, on an unusually cold day when the wildflowers over Prim's grave shriveled up, turned brown then surrendered their pedals to the winds. He was as terrified of not having a leader as the two brothers had been afraid of being degraded by Thel's tongue or clammed by his fist when they were all boys vying to see who would become men first. "Prim was our leader. The boss that the hunting, raiding, defending, farming, fishing, and building bosses all answered to. Who will be our bosses now?"

"Ourselves," Grim suggested.

"And each other," Dral added. "Everyone gives according to their needs, gives according to their abilities. And the one with the most abilities becomes—"

"—Someone more clever than you two masochistic, self-deluded idiots," Grim and Dral heard the voice of a stranger entering Prim's private hut. The visitor was clad in soft, shiny furs, which blinded their bloodshot eyes with a strange color that was a mixture of blue and red. It was a hue on clothing that no Primtonian had ever seen. Clothing that lacked leather and hair. "You know that the peasants where I come from live far better than this king does," the visitor said pulling down its hood, revealing the face of a woman. "With holes in the walls that you can see through and open and close as you want or need to. And floors that are so clean that you can eat off them without tasting any dirty, or sawdust. And old people where I live—"

"—we know, live longer than any elder here, by as many as ten winters," Dral said to the visitor from the settlement they could never find.

"Twenty," returned the visitor with no hair on her face, legs, arms, or back. And no blisters on her hands. And a face which was NOT that of the Huntress Grim and Dral fell in love with and somehow recognized. No, this woman was...as mean and vicious as she was beautiful. "But this ugly and primitive 'settlement' has its charms, I suppose," she said with an upward chin and condescending grin and eyes that fired into Dral and Grim's brains. Her stare blasted holes in their sense of self-worth in perfect harmony with every comment she made regarding how far they had 'advanced' from 'crude necessity' to 'common mediocrity'. "It's time that someone other than your pride-seeking and overly protective father, who is gone now, give you some criticism that isn't coated with honey or anything. Follow me if you deluded and idiotic cowards dare to, of course." She walked outside of the hut towards the 'common area' of the settlement, drawing villagers out of every hut, house, and hiding place in the bush to gather around her.

It felt like an offer that couldn't be refused. And if it was, would Grim and Dral's skills in 'magic' disappear as quickly as the summer and fall plumage were being blown away with oncoming winter winds? As potential and unwilling political rulers of Primton, they now had something to lose. Both brothers feared, above all, that they would be demoted to being… ordinary citizens by the visitor. Besides, it was personal with her, whoever she was.

NO one had called Grim and Dral idiots or cowards since they came into camp riding Prometheus' big dogs, who perhaps would also shit on them if they didn't stand up to this new blue-red-clad visitor.

The visitor insulted every one of Dral's inventions, as well as advancing medical procedures. Then she made fun of Grim's jokes, music as well as his stories. It was not like the Huntress, who merely said that things were better where she came from, showing respect for Grim and Dral and everyone else in the village. No, this intruder clad in cloth that the village had never seen, in a color she called purpose, emanated active disrespect. She tore down every Vision Dral and Grim had made possible to be seen, heard, and felt by the human eye, ear, and hand. Particularly the one they valued most.

The villagers seemed to agree with the Intruder. Just like in 'golden times of old', they looked at Grim and Dral like they were the losers. Losers they had been before meeting Prometheus and connecting to the Minds within their unused brains that he activated somehow.

Both brothers knew that three more insults from this intruder, who was so much more effective and colorful at criticizing than they had ever been at doing and would do, would end the dream and their life. A fate worse than becoming stew that would become shit.

A stroke of genius arose out of desperation as the Intruder, big I, moved towards Dral's best-constructed cart, upon which were mounted musical instruments to make the work of transporting heavy objects feel like play.

"You see what we've done here, with nothing to start with," Dral pointed out before the visitor's luscious, alluring, and fire-breathing mouth could utter yet another criticism based on fiction or fact. "What have you done to improve the lives of others?"

"With your own hands, head, and heart," Grim challenged. "On your own!"

"I…eh….know important people," she said, looking downward, having been hit in the area of her brain where self-esteem dwelled. "And have worked with very important people," she insisted, with a defensive voice that his remark somehow was able to evoke. "And I am very accomplished! And effortless success is the best kind!"

Grim pulled himself back, scratched his chin, and put his back into a hunchback position of a commoner. "Well," he humbly offered as the lowest commoner in any settlement, village, or campground, which he had grown up adopting and being pushed into. "I suppose that working under important people who have been given success rather than accomplishing it yourself is better than actually accomplishing something yourself... hmmm." He turned his brother Dral. "But what did we insignificant, small, and masochistic idiots on the bottom of the totem pole do?"

"We did things," Dral announced proudly, humbly, and accurately. "We accomplished things. On our own. Through struggle."

"Like every other commoner in this common settlement did and is doing," Grim proclaimed, referring the critic to the students he taught to be artists. Then, those who became innovative builders and healthcare workers by working with, and not under, his brother Dral. And finally, his sons, daughter, nephews, and nieces, in whom the seeds of struggle, wisdom, and intelligence had been planted.

"And as we all know, or should know, if we have any brains, balls or benevolent instincts to become better than we were created, or made," Dral added, somehow finding within himself the ability to use alliteration in a musical and accurate manner. "Effortless success is failure, not success."

"And there is no success like failure, which we turn into self-made accomplishment, which you, who are great because you are affiliated with great people, somewhere, will never know," Grim added. After which he came up with a song on the spot. The lyrics of that emotionally evoking tune described the Intruder, big I, as a pathetic soul who should be shunned, ignored, and pitied. It was accompanied by a new melody on the now triple octave lute, which, by the third verse, was backed up by other villagers playing the multitude of other instruments Grim had invented or improved. The rendition of that anthem was helped by Grim's linear sequential brother Dral, who now somehow found a way to play the scientifically enhanced flute, emitting music as well as merely notes.

The intruder left in disgrace. Grim and Dral felt victorious, for the moment anyway. The people they were in charge of now seemed to be liberated from and educated about, for the moment, this new invention that had come their way. The visitor who criticized everything so well but never did anything on her own felt a mirror forced into her face.

"Me and my people are not finished with you two idiots and the rest of you morons!" the well-dressed critic blasted out. She left, her head bowed low, feeling like she was a loser in rags. She stumbled back into the woods, slipping on her long purple royal robes, falling into the mud several times.

"So, Athena," Prometheus, as himself, said to the Intruder who had entered the village in a purple robe as a queen and came out as a raggedly unaccomplished loser. He put a real mirror into her face, which he angled into the sun so that it would reveal the wrinkles developing around her eyes, the enlargement of her once blemish-free and uninjured small nose, thinning but still long hair, and sagging breasts."Take a look at yourself!" he said, to the tune that Grim had devised after standing up to her. And exposing her to the world of earthling mortals and herself. "You were born with exceptional beauty to the most powerful and, so far anyway, father imaginable, and now you are... ugly and alone."

"As are you!" Athena blasted to her collaborator on the scouting Mission to Earth. Prometheus now tied her wrists in knots of unbreakable earth-made braided rope, which the mortals on this planet had learned to make from fragile green vines without his or her help, according to what she knew. But Athena did know that Promethous was just as ugly and alone as she was. With all of her might, nearly breaking her already painful arms, she pushed the mirror into his face. "You are ugly and alone," she said as the solo expedition leader who had been her lover back on her home planet had aged just as much as her. He now had more deformation on his potted, disease-speckled face than hers. But as Prometheus looked at himself in the mirror, he didn't seem to mind.

"Age is something nature created. And, Nature never gives you a problem without a solution," he noted, without a single breath of regret, nor tear of loss, nor fear-induced quiver of lips that bordered a mouth with more sores and lost teeth than hers.

"And the solution is?" Athena blasted out. "Back home! I fucking hope, soon!" she said, looking up to the sky.

"Or the solution is here in the meantime," he suggested, pointing Athena's attention toward the celebratory settlement. "Which you almost destroyed," he grunted. "Why?" he asked his fellow expedition member and once trusted and beloved lover.

"Orders," she blasted back, with tight lips and a closed heart.

"From your father? Zeus?" Prometheus inquired. "Or… someone else I'm not supposed to know about."

"Untie me from these ropes, and give me that transmitter you stole from me when I was sleeping, and I'll tell you," Athena replied. "Promise!" she pledged, putting her tied hands as upward as she could.

"Promise to what and who?" Prometheus challenged. "Ourselves, as gods who have to find a new planet to live after we finish destroying our own?"

"There are worse options," Athena answered. "Swearing to demons or to nothing at all."

"True enough," Prometheus replied. "But, I will let you go home and get 'better' under one condition."

"Which is?"

Prometheus thought about the answer. He didn't find one. At least that he revealed to Athena, for reasons he kept to himself, he cut the ropes keeping her confined, and gave her the transmitter. It was drained of most of its power due to climatic factors which Mother Nature provided without consulting with the Science Prometheus et al. knew about. "Get out of here if you can," he said. "Or if you dare to. You know what awaits you at home. And the possibilities for staying here."

Athena didn't answer. But Athena was not willing to leave to be resupplied with hopefully medicine on her home planet that would restore her. Or maybe not. Still, she needed one answer to one question. "Those implants you put into the thick heads of those two idiots, you turned into geniuses. What happens when they stop working?"

"They already have, a long time ago," the once confident space traveler confessed and related.

"How long ago?" Athena asked. "You mean these more-heart-than-brains humans on this primitive, uncomfortable planet are getting smarter by the day, on their own. Smarter than us?"

"And wiser than us, we hope and pray."

"Pray to who, what, and why?"

"We'll have to figure that out together," Prometheus replied. After which he looked at the watch on his prematurely arthritic wrist with a fear-induce shake. "Very, very soon!"

Athena perhaps heard what he said, but Prometheus knew she wasn't listening as she retreated into the woods with the transmitter unit. "Better that she takes the last portal home to where she could be restored to what she had been, if still possible," he thought to himself as his body felt another tooth fall out of his mouth, and his hand experienced pain upon trying to flex it in a fist. He was angered at himself for setting in motion something he had never fully told her about nor understood himself. And for letting the woman who he once loved learn to hate him so much. But, finally, he fulfilled his childhood wish to be a 'hero' who could perhaps save two worlds. But perhaps at the cost of everyone in both of them.

CHAPTER 9

Dral didn't only think differently than his technologically less advanced fellow humans. He now looked different. But did he now resemble the men, and perhaps some of the women, in the Huntress' advance settlement, which he still could not find or locate? Still, he persisted, this time without his overworked and more popular brother, down a new path through a recently formed valley after an unexpected avalanche that happened immediately after lights came down from the sky. While washing his face and filling his dry throat with a smooth, surfaced lake, he inadvertently helped himself to a view of himself. He noted a glare coming off the top of his head, seeing in his hand the last batch of hair that had formerly taken root atop his overworked cranium. It came with an active sensation that had started after meeting with Prometheus on the right side of his head and, after a time, shifted to some extent to his left. "So, no grass grows on busy streets," Dral said to himself, repeating the metaphorical remark that the still generously-follicled Grim somehow translated into a humanistic joke. "I suppose the lice will have to find someone else's head to move into," Dral said with a half smile on his left face. "And I can feel the thought waves from other beings and places coming into my head more easily than others," he continued, feeling both sides of his lips pulled upward.

His thirst for water quenched, and his required daily dose of self-observation administered, Dral got up on his feet, looked up to the grey sky, and grabbed hold of a magnetic dial which on a sun-less day or moonless night could tell him which way North was. It was where it had been when the sun had shown much earlier than morning by means of where it cast shadows. An essential tool to avoid becoming lost like had happened so many other times when he went too far away from the settlement, and he speculated something that could allow one to find one's way across and back from the lake that contained salty water and no visible land on the other side.

"So, where do we go from here?" he asked his horse, who by now was just as old in horse years as he felt he was in human age. "Do we collect the medical herbs from our secret mountain to the left before winter, animals or intruders we still haven't seen deplete whatever is still there? Or do we take the pass on the left to try, again, to find the Huntress' settlement where they have better medicines? Or maybe we'll try to find the people who took the plants we found and planted on the mountain that Prometheus said would never be depleted as long as we..."

Dral's discourse to the horse who he shared more secrets with than any human, including his brother Grim, was interrupted by a rustle from the woods behind him. The steed's ears went forward. Then his front hind limbs went up, hitting the ground in a turn that set him into a gallop faster than anything he had done when a younger horse. "So, NOW you show me that you're not lame!" Dral yelled out to the horse. "All of that limping you did when I asked you to do more than a slow trot was an act?"

Dral's nose picked up an odor that scared him as much as it terrified the horse, causing the steed to throw Dral off his back and then gallop away from the smell as quickly as it could. That fear-inducing aroma was the smell of rotting flesh in the bush. Its source started to move. Perhaps it was an injured wolf out for a last meal before he died. Or, Dral found himself imagining as real, as it was not impossible, the 'walking dead'. Dral braced his scientific mind for the latter with a quick review of the legends about such which were maybe based in fact. He attempted to protect himself from becoming one of the walking dead by picking up a long blade with a handle on it. That same instrument left by Prometheus, which could be used to cut thin branches of trees, the ribs of slain animals, or, in the wrong hands, the limbs of non-Primtonians who wanted to take over his homeland.

"I'm harmless, now," the purulent-smelling puss-covered beast said as it emerged from the woods through the fog. It showed itself to be a crawling creature that, with some effort, was able to walk on its two hind limbs. It revealed itself to be covered with blood and purple rags stuck to flesh under it. As it got closer, it showed itself to be a woman of more advanced age than any Dral had seen. A very elderly one who, for reasons beyond her control, still had not surrendered to death. "Can you fix this?" she asked through a deformed mouth, spitting out as many desperate words as blood.

"I...eh...will try," Dral said as he reached into his medical kit, thinking about, as he had to on rare occasions, putting her out of her biological misery with an overdose of sleeping medication.

"I was talking about this!" she continued, with forced words through quivering lips. She pulled a device out of what had been her 'dress' with two shaking, emaciated arms that she somehow made work together. It was coated with blood, the sharp wires sticking out of it coated with blood. "A beast got hold of it just as I was about to...eh..." she said with fear and regret in her voice as she became faint.

Dral reflexely broke her fall and assured her fearful soul with a bear-like hug. He recognized the old woman's eyes but still held on to her. Indeed, she was the once beautiful and arrogant intruder who introduced Primton to its first highly skilled self-respect-demolishing critic. Some kind of ailment had taken away her youth beauty and was about to take away her life. But Dral had to help her. And, he thought, if he could keep her at least live long enough, she could take him to the Huntress' settlement. And, perhaps, if Fate was kind to the Huntress.

"Lay down here. I can help you," he said to the ugly, aged version of the Intruder Critic who, days earlier, had been a picture of youthful beauty. "First, we get rid of the pain safely," he said, helping her lay down on the ground without falling down onto it with a thud.

Dral reached into his medical saddlebag, which he had, thankfully, taken off the horse and thrown upon his shoulders before it ran away. After administering what he thought and calculated was the right dose of elixir, he felt her skin at the diagnostic points, which, in his experience, were soft and shallow when associated with specific maladies under the skin. "I think I can save you," he said. "But I'm going to need you to believe that what I'm doing will work," he said, having verified that 37 out of every hundred patients who merely believed that he was a brilliant healer did in some way become healed without him doing anything.

Before he gave the aged and soon-to-be-dead woman the medicines that he knew would work, he palpated her swollen abdomen. "Yes," he said after examining her chest, legs, and aching head, envisioning what was happening under such. "I know that I can help you." He instructed her to open her mouth, placing a pinch of specially formulated powder under her tongue. Her pulse got stronger, her pale white face acquiring color again, her white gums becoming pink, her breath moving air in and out with a now diminished death rattle. "Yes, that's a start," he said to her. "And when that horse of mine comes back here, I can get you to a place where I can do a lot more…"

"Which you can do better if you fix this first," she said, pointing to a metallic device strapped to her waist. "My hands are too mangled to move any of the wires, to put them in the right place. And my head…can't remember where all of the wires are supposed to connect," she related, demonstrating the truth of those observations. "Please, fix this, and you can fix me, you, and many others," she claimed, her chest, arms, and mouth spewing out blood.

It wasn't brain surgery, putting the device back together. By application of simple logic and mechanical intuition and forcefulness of fingers, Dral was able to put the metallic tool he knew nothing about together again. Just as he was about to figure out how it worked, the old and not yet dead woman maneuvered her now non-shaking hands such that she could hold onto the device.

With a religious reverence, she pointed it at Dral's head as he slapped on her chest, arms, and into her nostrils an extract from a plant he would name after his father, brother, or perhaps the huntress. The wounds he could find stopped the bleeding. The old woman's pale complexion regained color. Her pulse became regular again and strong. Light returned to her eyes.

As the old woman chanted what felt like a death song, Dral could feel something light up and vibrate his right brain. He could feel the seeds of ideas yet to be understood planted and germinating there. His fingers knew exactly where to insert splinters of wood into the skin of the old woman, which he twisted to the point that they were cold and shallow, 'gobbled up' wooden 'needles'. He felt more electrical shocks going up his arm than when he normally did this manipulation. They were amplified when he attached the metal wires in the metallic device to the tips of the needles.

The old woman then pointed to the device Dral had repaired, instructing, then begging him to lay it on her chest.

After obeying the old woman's request, Dral looked up from the body he was healing to the soul who owned it by looking at the face of the woman. She became younger and... thankful. "So, you think we can become partners," the old, dying, then middle-aged healthy, then youthfully attractive woman said. She placed her now functional, non-bleeding, and non-purulent hand on his shaking forearm, then startled face. "Yes, partners," she said, as a character described in Grim's legends and songs. An enchantress, healer, or 'goddess', or some combination of such.

"I...eh...have a wife who needs me," Dral confessed and related.

"But who doesn't love or understand you," replied the miraculously restored female critic who had to be from a place more advanced than Primton. She now showed Dral respect and, so it seemed love. "I can give you what you need and want and can give to those who you give what they need and want. Starting with…"

Dral took in a deep breath, preparing himself to be pleasured between the legs in the way that his lesser-evolved patients lived, fought, and were willing to die for. Instead, he felt the now repaired, upgraded, and light-emitting device pointed at the right side of his head. Whatever it fired out caused him to feel on fire inside his skull, and a thunderbolt went up and down his spine. Before he could figure out some biological explanation or envision a medical fairy tale as to what was going on, he fainted.

Soon afterward, Dral woke up from a strange dream he couldn't remember by his horse. He heard the old woman's chanting as a young one. Her voice seemed to come from the woods to the North, South, East, and West all at once.

But there was one voice that did come from one clear direction. With a voice, he knew very well. "Are you okay?" Grim said, one of his hands holding the reins of his and his brother's horse. The other was on his brother Dral's shoulder.

"How did you find me?" Dral asked his brother.

"The question is WHY," Grim replied. "We're both needed back home. You more than me."

With that, Dral and Grim galloped back to their home camp, which, now, for better or worse, was a town.

CHAPTER 10

The horses could smell the disease in Primton from a mile away, courtesy of a wind that blew its aroma into their nostrils. It set their ears forward and their legs into convoluted choppy gaits that would normally land their riders onto the ground with broken forelimbs and hindlimbs. Thankfully Dral figured out what salve to put into the horses' noses so that they thought they were returning home to fresh, springtime grass rather than the stench of liquid blood-stained human feces and the fragrant aroma of vomit coming out of the anal orifice. Upon arrival, Grim's ears heard something even worse.

"Death rattles," Dral said of the people who, when he left, were merely uncomfortable in the belly and now sprawled on the ground awaiting exit to an existence which is, according to the stories Grim felt obligated to invent for them, kinder than the one they were born into.

"I'm afraid we'll lose more than just the old and feeble with this one," Grim added. "Because Mother Nature this time—"

"—is an irrational creature who is just as trainable as last time," Dral blurted out of his mouth as he smelled the excretions lying around the sick and dying. After which he scratched his chin, then raised his index finger up into the air, "An interesting disease this time which requires an interesting remedy." He began concocting another combination of powders taken from 5 of the bags in his cart. "An interesting experiment which I am sure will work," he said with an 'emotion' far more different than his usual tolerated and expected non-musical arrogance.

"Detachment," Grim said to his confused horse, who was alarmed at the stench and sounds coming from the sick villagers. "Detachment from the lower emotions that make Dral feel more than think. Detachment from letting empathy for his patient's pain interfere with figuring out how to stop it. Detachment from whatever musicality is implantable into his soul. And…" yes, Grim had to admit to himself and share with the horses. "Detachment from me and you as living beings rather than tools to do his self-assigned duties and Callings."

But that sacrifice of Soul seemed to be necessary for Dral to emerge from being a medical miracle worker to being a god. One who indeed had divine knowledge about the Natural World in his head and magical fingers to administer that biological wisdom. Somehow, Dral was able to save ALL of the patients from dying that day. Somehow, he converted all of the sick into being healthy within hours of his tailor-made treatments. It was through means that he didn't bother to share with any of the people he saved.

He wanted to share his medical wizardly with his brother, but doing such would involve relating it in a medical language Grim did not understand. Dral was unwilling to and, when pressed to do so, unable to 'humanize' his technical explanations for his left-brain artistic brother or any other art-loving soul in Primton. Though he seemed to become a master of biological science relative to what he had been that morning, there was a deadness in his eyes. And a flatness in his voice, to the point that every word in each sentence was delivered with the same volume and intensity. It was something that had a calming and sedating effect on anyone who listened to it long enough. Something that made the always needing to feel more Alive each moment artistically Grim feel...

dull. Like he was a saver and giver of life who was, in the ways that matter, dead himself.

The next day, Dral had figured out what caused the shitting and barfing disease. It was an infestation of 'insect-like organisms smaller than any that can be seen by the human ocular senses' that found its way into the well from which the richest Primtonians drank most and the poorest when they could. He then drew out in the sand, and onto sheets of thin wood that he now called paper, what had to be done to fix the well. The process of 'disinfecting the well', made it fill up with water faster and easier. The day after that, Dral drew out a plan whereby the water from the well could go into each hut through hollowed-out logs as needed, as well as into the fields where the villagers had been taught to grow food.

A week or two later, Primton had more water and food than its citizens needed. A month later, as measured by the cycles of the moon, later, there wasn't a thin citizen in Primton. A month after that everyone in Primton was comfortable. Some were more comfortable than others. Comfortable in their bodies and... tragically... souls. And in their interpersonal interactions.

But their lives lacked humor. Grim knew and informed, people that without a joke offending someone, somewhere, sometime, it was not funny. Emotionally offending (or challenging) others, past, present, or future, became a legal offense in Primton, punishable by taking away comfort-conferring food, clothing, lodging, and medicine. Also, saying anything that was considered offensive or disturbing by ANYbody was punishable by taking away various comfort conferring things inducing pain in the body. Or, worse, the worst punishment imaginable for the lesser evolved citizens of Primton— banishment. Banishment in total or in stages.

Grim was asked, then required, to write songs and stories that lacked any offensive material, most importantly the soul-challenging and perspective biologically unexplainable phenomenon called humor. Humor was always the hardest thing to put into a singable song and performed story and the first thing to come out when the story or song had to be given sufficient structure to be understandable and producible, even for Grim. His students learned all too well how to take it out of stories and songs. It left Grim more alone than he had ever been. He was the only Primtonian non-afflicted with Dull Out Disease. The order-producing disorder made people and what they produced lifeless, boring, procedural, safe, and...comfortable.

Then, the inevitable happened, in stories that Grim saw in his still expanding head, the contents of which he hid with a closed mouth as well as heavy skull-covering hats. It materialized in the world he could see all too clearly with his biological eyes, or as Dral and the technologically advancing soul dead citizenry who considered him a god would term them, 'ocular portholes'.

CHAPTER 11

Primton still kept its original name as the new rulers saw fit to honor the great old departed warrior Prim rather than diving with BOTH feet into wars with each other. No one with any real brains wanted the job of being king in a democracy where the king was replaced almost as often as there was a new full moon. And, to be truthful, anyone who wanted the job proved to not be trustable with it.

But, no one did anything against Dral's, as he now called it, scientific recommendations, as they were afraid they would lose their heads or, worse, popularity if they did. And, of course, every ruler boasted about how he or she was Grim's best buddy and mentor before he became a genius artist.

Grim allowed those fables to be accepted as fact because the rulers now had an army of muscle to back them up. They were far larger and unlike the 'band of three' headed by Thel in the legendary 'good old days' when Prim was the leader of a small band of hunter-gatherers who somehow wound up surviving the perils of being killed by Mother Nature. Or eaten by wild animals, some of those beasts being their fellow two-legged humanoids from other valleys.

But things were 'looking upward', according to Klep, the newest good-looking man-king who knew how to charm the crowd with what they wanted to hear rather than what they should know about. The population of Primton went up to numbers that shot into 'as many as the stars' due to increased birthrate amongst the native Primtonians. Most of the babies popped out of those who were well-endowed between their legs rather than between their ears. And, of course, Primton was infiltrated by 'lesser' but still needed people from other regions. Those 'barbarians' were conquered by the spear, then arrow, then metal sword. Some were put into submission by a weapon more powerful than any instrument designed to tear apart human or animal flesh. It was a new system that defined who was rich and who was poor.

"So, who decided that this shell coated with red dust is more valuable than this pebble coated with silver muck?" a young boy eleven winters old asked his mother while sitting on a piece of wood that supported his back and ass inside his rebuilt hut, which was now a house.

"And who said that this dark yellow nugget is worth 100 red-covered shells and only 20 silver muck-coated pebbles?" his younger sister pressed, looking at the gifts her Uncle's assistants had delivered to the elevated board in front of them, which they now knew as a table.

"And who said that a handful of five yellow nuggets is worth more than a herd of horses or a person from somewhere else we don't know or somewhere?" the boy inquired of his father. He dipped his not-yet hairy hands into the basket in front of him filled with shiny objects that one couldn't eat but were necessary now to obtain food or most anything else.

"He or she who has the most yellow nuggets says how valuable they are and makes the rules for everyone else," a very comfortably, moderately expensively, and fashionably beautifully clad Lilila informed her children. "Which is us, because you both are—"

"Grim's children?" the boy exclaimed.

"No," the not-so-young-anymore mother Lilila informed the still young and still happy for the right reasons children. "We have more yellow nuggets than almost anyone else because you are Dral's nephews and nieces."

"And who has the most yellow nuggets and can buy anything or anyone?" the girl inquired.

"The king, of course," the boy informed his little sister.

"But the king is rich because of what our father Grim used to do a lot and what our Uncle Dral is doing a lot now," the girl noted, lamenting the former and proud of the latter.

Such aroused a suspicious eye roll from the girl's now jealous brother, who was more his father's son than his mother's. He was angered at his father for, as his mother said, 'was spending too much time taking care of other families and imaginary ones yet to be born in the world inside his head'.

Lilila could smell another fight developing between her two children. They were the only children that Grim was able to produce for her. And now, they were the only two lives, other than her own, she really cared about. She pulled the soon-to-be sparring for superior children into her with a big hug.

"Yes, the king and his family are fatter than us," Lilila admitted with anger that she tried to hide. "They have a bigger house. And more comfortable warmer clothes. They are rich in gold nuggets and possessions. But they are not richer in Vision, according to your father Grim and, more importantly, your Uncle Dral."

"Who do you wish could be our father and your husband?" Lilila's daughter asked in an angelic voice.

"Which is what we both heard you pray for to the lighted ghosts above the clouds," the boy informed his mother. "And if the ghosts give you what you want and what you say we need, what would have to happen to your sister, Rihi?"

Lilila turned around to her son, seeing in his not-yet hair-covered face the eyes of a young and soon-to-be cynical man. One who she hoped she would not have to silence with a slap across the face at home or a spear into his heart on a walkabout into the woods. Yes, to be the wife of the 'down to earth scientist' Dral, who was now more prestigious in Primton than to be married to his brother, 'head in the clouds artsie' Grim ever was. Such would be good for her children and for her. And, she told herself, perhaps better for Rihi's four children she had by Dral. Or who she claimed were had by Dral anyway, despite the fact that only one of them resembled their father. And one of them seemed to be the spitting image of Grim.

But feuds between sisters in law fought presumably for the welfare of their children would have to wait. A burst of wind blew into Lilila's house, followed by rocks from other huts blowing small openings in the wall. Then boulders, making king-sized holes in them. They revealed a full view of the large fields below of special crops developed by Dral that, he claimed anyway, would provide sufficient food to enable every man and woman in Primton, be they owners of gold or property of owners of gold, to be as healthy, fat and comfortable in the belly as the king.

The wind blew into the sky almost every stalk of remaining 'wonder plants' containing all the human body needs in sickness or health if eaten in the right proportions. Dral's latest, and some say greatest invention, was now needed to replace the wild game now gone from anywhere within five days' walk. Or two days' wagon ride from the now overly populated 'kingdom'. Such was due to overhunting as well as the destruction of the forest to build houses for people in town.

A wave of insects swooped in to demolish what they could of the previously harvestable abundant and now preciously needed crops in the fields guarded by heavily armed men who were helpless to fend off Mother Nature reclaiming what was hers. The wave of wonder plant-eating insects was swiftly followed by thunderbolts from above the clouds, which set the fields ablaze.

Lilila grabbed hold of her children, cursing Grim for not being there to help her. She assured them that everything would be alright. Such registered as fact with her daughter, but a lie to her son. A boy who, if he had too much of his father's wit and will, would grow up to be an enemy Lilila would have to silence. Or, if animal flesh and wonder plants were not available, a slab of meat she would perhaps have to eat.

Lilila knew that 'wonder brothers' Grim and/or Dral would have to work some 'angelic magic' to restore her home and life to its high standard of living. Or, she pondered, perhaps it was Grim and Dral's black magic under the control of demons that enabled their fellow Primtonians, often at the expense of 'barbarians' from elsewhere, to become so comfortable and happy. Lilila considered that there was a moral accountability to it all. Maybe Primtonians had been set up to know comfort and security so they could experience even more intensely pain, misery, and terror. In any case, the final trial to see if Grim and Dral were angels or demons would come very, very soon. With a congregation of commoners on hand to elevate Lilila and her sister Rihi to queens again. Or, perhaps, a mob that would burn them at the stake along with the two downtrodden losers who had cheated their way to becoming temporary 'winners' in the battle between mortals and Fate.

CHAPTER 12

"Nature never gives you a problem that science can't solve," Dral insisted to the congregation of native and foreign-born Primtonians when they found their way back to the settlement after the multiple storms that left no citizens or slaves without some loss of property or bodily function. He strolled around the center square of the settlement, perusing what remained at the plateau under his feet and the once-forested valleys below. Then he looked at, but not into, the faces of the people who called it home. "Science enabled us to come out of the stone age before and will enable us to do it again. All biological systems and civic experiments, if maintained, always get better. And all communities can be resurrected from even the worse setback." he proclaimed.

"As long as they are well motivated," Grim pointed out, stepping into the speaker's circle from the aloof log stump that had been the center of the once vibrant outdoor 'performance stage' which made real-life bearable, understandable, and transformable into something worth living for.

"And things will return to normal if we are all fed," Lilila shouted out, disgusted and ashamed of her husband. "With something beyond just dreams."

"For the moment, you are correct," Grim conceded, feeling the grumbling in his own stomach while seeing the hunger for physiological as well as metaphysical sustenance in everyone. Such included his own beloved potentially multi-perspective son and his, on a good day, tolerated uni-dimensional 'meat and potatoes' (and once sought after above all others) wife. "Dral?" he said, turning to his contemplative, follically challenged brother while picking up a lute, which miraculously had survived the unpredicted storms from sources unknown to any right or left brain genius. "No grass grows on busy streets, but there's a fire in there that I'm sure you can use to pull something out of your ass. Or you can fart out an answer to Mama Nature and Father Fate to make them send down apologetic milk, honey, and bread from the sky instead of pissing rain and farting out more wind," he sang. It was admittedly with lyrics that could have had more edge, but it somehow got some frowns turned into smiles. Most importantly, in Dral, who rose to his feet.

"If the Fates don't want us to be alive…we will create life!" Dral promised the people and yelled up to the sky. "The sun is still here, and, most of the time anyway, is still our friend," he announced. He then took in a deep breath. "And the air we breathe in and out is still here. Even though some claim that the air coming out of overinflated nostrils from thick-headed scientists makes breathing in less fun for others," he continued, looking at his brother.

Grim smiled in delight, admiring his brother's attempt to activate the humor part of the left side of his head. A third brain seemed to be developing between them, yet again. But more importantly, this time, it was a third heart as well.

"So, what do we 'mortals' who live in the real world do now?" Rihi asked her husband, speaking for the rest of the congregation who were at the mercy of hunger and, according to the thick coats on the horses and dogs about to be eaten soon, early winter.

"You think you are a god who can create life, 'Doctor' Dral?" Lilila pressed.

"No, something more powerful than a god, a determined man," Dral replied.

"And, a man who, yes, has connected to the woman inside of him," Grim added, finally admitting to whoever was curious or listening one of the secrets behind his magical ability to create humor, song, and story that related to both genders.

With that, Dral drew in the ash-covered ground the plan for creating life. "A photosynthesis machine," he said, feeling a concept from the future come yet again to the present. "Light comes into this portion of this box on top. We breathe into the holes on the sides of it, giving the plants to carbon they need, and out the other end, a plant emerges! From the seeds we plant into this dirt."

"When?" still King Klep asked, on behalf of the people who he was only marginally smarter than. Such was the reason for him being their leader who would never advance beyond that position. "And who will pay for this 'miracle'?" he pressed.

"Whoever wants to," Dral announced.

"And needs to, which is now…everybody!" Grim reminded the congregation of haves and have-nots, not knowing who now was in which category. "We are all in this together now," he said, after which he strummed his lute again. The lyrics of yet another motivational song about to come out of his mouth were halted by Klep's yapper.

"And if this photosynthesis machine fails?" Klep said.

"It won't fail!" Dral blurted out with assurance.

"Shouldn't," Grim added.

"And if it doesn't, you can feast on OUR flesh first!" Dral added.

"Which we will," Dral and Grim heard with their ears and saw in Klep's eyes. And, for the first time, in everyone else's. They understood now more than ever that the gifts they possessed to serve people (who they now outgrew being around and with) were both a curse and a blessing.

CHAPTER 13

"So, maybe these advancing primates will pull it off this time," Prometheus noted to Athena as he viewed the experiment on the lower plateau of the mountain now named 'Primtonia' from the rocky overlook above it created by Mother Nature on Earth, or perhaps on purpose, by Zeus' second expeditionary Starfleet. Or perhaps the reshaping of the terrain was a collaboration between the 'Earth Mother' and the humanoids on his home planet. Or maybe all of these unpredictable changes creating new mountains and the variety of species to live and die in them were initiated by expeditionary humanoids from Prometheus' enemy planet who he did not know about. "Creating life, even plants, from inorganic sources, could be innovative," Prometheus noted.

"And dangerous. And, for these mortals, at this stage, not allowed," Athena retorted.

"By whose rules?" Prometheus pressed.

"The ones who will see that you will pay for what you've done here, in the service of your ego," Athena shot at and into Prometheus' defiant face, leaning back on a hard rock that she converted into a soft pillow using whatever battery power was left on her transformer. "And that book you want to write to advance your position in the Academy back home."

"An academy which needs the help of these primitive mortals more than you, I, or anyone else, especially them, realize. Particularly when we erroneously decided that it was cheaper to conquer our sister planet back in our own galaxy to get the raw materials and manufactured goods we want and need, instead of paying retail prices for them," the explorer from the planet where humanoids were three-quarters brain and one-quarter heart said of the planet where it was the other way around. "One look at either of us on the inside and outside verifies that," he said, looking at his own still shriveled and weak hand infected by agents that could be maybe killed by the right herb. He then treated himself to a glance at Athena's prematurely tired and yellow-tinged eyes. "You really should consider some medication for that liver disease which you acquired on this trip, from a saboteur at home or a bug here. Which—"

"—yeah, I know," Athena shot back. "Makes me feel depressed and angry."

"And confused if it gets too bad," Prometheus replied, after which he picked up a wooden earth mug filled with fermenting berry juice. "A bad design of Nature or whoever was here before us that ripe fruit keeps brains and bodies alive, and rotting fruit destroys both. Starting with the liver. Which... I'm told, can grow back even after it's damaged by toxins or...."

A crow, coming in from nowhere yet again, landed next to Prometheus. He smiled, offering the bird a small drink of the berry juice, which was accepted. He then looked into the eyes of the black messenger, with the gut feeling that it was no coincidence that this bird from the lower altitudes below the clouds had flown up to him. And that the crow decided to visit him while he was talking about livers, destinies, and prices to be paid for doing good, noble, and experimentally expansive deeds. "But there is one thing I want, and need, for you to tell me, Athena," old, on the outside anyway, Prometheus asked his old on the inside and young on the outside fellow explorer. "Why did you give Dral an extra jolt of insight? With some brain implants that you stole from my medical kit?"

"So he could cure me of my biological problems," she replied, for the first time in a long time, honestly. "He's the scientist in this experiment of ours, right? And there is no such thing as too much science. Right?"

"As long as there is enough art, which is a more effective channel for humility and humanity, compassion," Prometheus replied. "Effective compassion is the ultimate result of intelligence, and ignorance is the only cause for cruelty." He waited for a response. Athena, after drinking a large portion of the berry juice in her mug, which was more fermented than what was in his, smiled, looked at Prometheus, and then slapped him in the face. She kissed him, then passed out in his arms.

Prometheus felt Athena's heart beating as it lay next to his. But beating even louder between his ears was the drum of Fate, with a rhythm of its own, to a song that would change the universe, about to be composed by the lowest humanoids dwelling within it. Such would, rightly or wrongly, alter the Fate of an advanced planet many light-years away from his own. A planet that would, perhaps, not know what a light year was for a long, long time.

Athena went into a kinetic drunken slumber. Her mind was taken over by the 'goddess' of sloth and sloppy loving. Her possessed fingers caressed every erotic spot on Prometheus she knew about when they were younger and secret ones she never knew about.

Yes, Prometheus' penis did turn from flaccid to erect. He kept hitting it to make it behave. He knew all too well the price of giving in to lower urges, such as Zeus having way too many children than he could handle or wanted. Those children had created grandchildren who he tortured, discarded, or, worse, turned into obedient copies of himself as enforcers of his will and promoters of his legacy. Finally, Prometheus stopped feeling the urge to dive with a smiling, happy face into the abyss the demoness possessing Athena wanted them both to fall into. He pulled away from Athena, laying her down on the rock she had converted into a soft pillow.

Athena's alluring love moans turned into ugly snores as she finally went to sleep. After being sure she did, Prometheus reached under her robes and retrieved what he wanted from her in the first place.

"Yes, interesting what this can do and can't do now," he said, examining the chronically breaking down the multi-purpose transmitter, which had been repaired by Dral. After Athena had zapped his brain with an extra jolt of bioelectric elixir that had enabled the earth born caveman to become a technical genius with intelligence levels rivaling the even above average humanoids on Prometheus' own planet. He checked each of the functions on the device and finally found that it could send him and Athena back to his home planet.

But the ability of the device to turn a two-digit IQ caveman into a four-digit genius, or a four-digit IQ ET 'alien' such as himself into a five-digit Sage, was now neutralized, locked into dysfunctional mode. It was restorable only by geniuses and devices on Prometheus' home planet. Zeus, of course, had forbidden it from being sent to the planet Prometheus was still standing on, as the 'Godfather' of the gods intended to colonize it with what was left of his fellow 'superior' humanoids. But, as for who was superior and what was inferior, Mother Earth sent a small reminder of Prometheus' way.

A cockroach made its way up Prometheus' arm, onto his hands, and into the transmitter. The insect looked at him and seemed to smile. It lifted its back end and emitted a drop of excrement into a main connecting wire on the transporter portion of the unit. The droplet of insect-derived feces and urine was just enough to enable the device to transport two other passengers, aside from Prometheus and Athena, to get a ride back to their home planet. Assuming that said home planet had not destroyed itself yet.

It was a philosophical moment and spiritual revelation of sorts. "So," Prometheus said the lone roach. "You haven't evolved in a million years and will probably not have to evolve for another million years to keep your place on the bottom. And, in your own time, like your other small pals who demolished the wonder crops, maybe one day YOU will be the highest life form on this planet. You so-called lower life forms, some of which we see and some we can't or don't see, remind us that we higher life forms can be brought down by you at any time. But as for letting us go from dust to dust and birth to being eaten by your progeny, we 'higher' life forms still have to try. Even if we have to create other life to do it."

With that, Prometheus smiled, looked away from the roach, then quickly made a swatting action with his hand towards its arrogant, smart assed compound eye-bearing face. Such succeeded only in making his aching flesh even more painful as the roach slithered away into the dirt. It left the rebel and too human for his own good 'god' to hope, yes pray, that Dral could create plant life to feed the over-populated and now, (after a series of 'accidental' earthquakes, volcanos, and other mountain moving geological events) geographically isolated kingdom Primton. Maybe the primitive humans Dral would save would evolve into a kind, Alive, and albeit colorfully dysfunctional sub-species that would not destroy each other's bodies, minds, or spirits. Like Prometheus' ancestors did once and were about to do again if this experiment of his failed.

The time frame for what to do and how to do it intensified, particularly when Prometheus saw, smelled then heard a flash in the sky that was something far worse than Zeus sending another 'follow my orders and no one else, or else' memo to his hopefully still favorite explorer. Indeed, it was a craft from Prometheus' friendly and now sinister sister planet, which was about to send down its own explorers.

Or, perhaps, it would reinforce them.

CHAPTER 14

Dral's photosynthesis machine was able to produce food that had all the essential ingredients. But when the human palate contacted it, even if it was connected to an empty stomach, it lacked all flavor. Dral also came up with an enzyme that enabled people to eat wild grass instead of barfing it up. It enabled them to grow healthy muscle, bones, and organs like the horses and other grazing animals could. But the taste of grass to people who he worked so hard to serve didn't please anyone.

The people, who had become used to being comfortable and obsessed with becoming more comfortable than their neighbors, demanded something else from Dral. Particularly after he designed and helped them build winter dwellings that were all the same, lacking any style or the comforts they had gotten used to, including rooms with heated seats where they could evacuate their bowls. Dral insisted that people other than himself come up with innovations to restore the now surrounded by strange new mountains, the Primtonia Valley, into a flatland kingdom again.

Grim defined, with Dral's help, the formula for making music that makes people think rather than decide it was more comfortable not to and employed it. The music with ever-changing tempos, keys, and rhythms could carry stories with thought-provoking twists and turns. Tales that would give the people what they needed to think their way into a better life, which was filled with challenge-inducing. Bliss infused 'dissatisfaction' with how things were so they could make them into how they should be.

But the people wanted something that calmed their terrified minds into passivity rather than challenge them into action. Something with a steady beat that never changed. They yearned for stories that told them lies that would comfort them rather than truths about how the world was and could be that would set them free from their self-limiting lives. Grim's works, particularly when he was bold and bright enough to put humor in them, were accused of being 'jarring', 'offensive', and finally, 'criminal'.

New weapons that Dral did not build with his hands and Grim did not envision in his stories found their way into the hands of the least intelligent (and, as a result, the least caring) Primtonians. How and why, neither of the brothers could figure out. But they did know that something had been 'undone'. Particularly when they were both brought into the center of the village, they had turned into a city, facing their new judge and new king.

The new monarch rose to power by acquiring more weapons, gold, and followers than anyone else, though that baffled both genius brothers. "So, why should we keep you short, weak, and head in the clouds losers around?" a seemingly taller and certainly royally clad Thel asked Grim and Dral from the center of the stage, which both brothers had used to address their former worshippers. "You are useless to us now."

"And dangerous," Rihi added from the queen's stump to the left of Thel, directing fire from her fear-fueled eyes at her former husband, Dral. "With the mark of the demon on top of his head that caused the hair to fall out! Who is… dangerous!"

"And his brother," 'queen on the right side' Lilila said with a condescending eye roll to the now family-less Grim. "Someone to be laughed at rather than laughed with," she continued with an all-knowing smile. Her finger pointed at Grim's functional but not fashionable attire. It was followed by a remark that Grim didn't understand, which evoked ridiculing laughter from almost everyone in the crowd directed at him, except for his son, who hid what he was feeling and planning with pulled-in lips.

Grim's son's face was so expressionless that it had to belong to someone with conflicting emotions of the most intense kind. There were three others who didn't join in the laughter directed AT Grim, who the demoted artistic genius didn't recognize or was able to otherwise identify by age or gender. They put hoods over their faces, then slipped into the woods.

Rihi pointed to Dral, accusing him of being a word in some strange language that meant something to be hated more than ridiculed. Such evoked boos from the congregation, which was now a mob. A wave of hurled shit and rotten artificially-produced 'wonder plants' added another coat of brown and red on the bruises inflicted upon them by the new 'law and order enforcers' who had pulled

130

Dral out of his now destroyed laboratory and Grim from his self-produced 'library'. Both facilities were now burnt ashes.

"What is our crime?" Dral demanded to know.

"Except for giving you people what you need instead of everything you want?" Grim yelled out.

"And trying to serve rather than please," Dral declared.

"By struggling to be useful instead of just doing what's easiest for us to do," Grim added.

"There you go," Thel said, with a voice indicating a cunning that was far in advance of his intelligence or normal wit. "Accused from their own blasphemous mouths! Thinking themselves to be gods when we know they aren't. And the gods know they aren't!" the all muscle, no brains, and more handsome than ever Neanderthal leader continued to his people with a royal presence and social intelligence, which both brothers sensed were put there by someone else. Someone who Dral and Grim could sense being around the valley. Someone with sinister agendas far more deadly than anything the Critic Intruder ever intended. An agenda that was too illogical to be initiated by Prometheus. Theoretically, anyway.

While being booed and ridiculed, Grim and Dral looked at each other, trying to figure out with the third Soul between them what to do. After reaching their conclusion, they metaphorically flipped what was now known as a coin. Grim won the toss.

"We consent to the worst punishment you can give us!" Grim, the left-sided artsy brother, proclaimed with the assuredness of a true scientist.

"As shits who should be treated like shit," Dral said-sung, on key, to the tune of one of Grim's earliest and now most hated compositions.

"We should be ordered to go far away from here," Grim said as he grabbed hold of his horse just as it was about to be turned into meat. "With these shitters, whose meat and excrements are toxic and cursed!" he told Thel's butchers with a spooky voice, scaring them into the woods.

"And never come back, to be at the mercy of the elements," Dral added as his horse came up to him, now released from the handsomely clad bold 'soldiers' who deserted their posts. "And we should die an ugly death!" Dral proclaimed as he got on top of the horse.

"To be eaten by these beasts," Grim said as he pulled himself on top of his steed. "And to die in the worst way imaginable."

"Which is to die... alone," Dral declared.

The two brothers rode past the pile of wood, which was to be their final, painful resting place, and disappeared into the woods. What they were feeling, they didn't know. As to what would happen to the families, friends, and patients who the mind, body, and spirit doctors were leaving behind, their minds didn't care. But their hearts somehow did. Yet one thing was certain in their lives that was not certain anymore. They had become a different species than what they used to be. The last of their kind. Or, perhaps, the first of their kind if they could find the Huntress. Instead, they ran into someone even more loving, caring, and intelligent.

CHAPTER 15

"So," Dral said to Prometheus after 'accidentally' bumping into his one-man campground. He was offered some well-cooked and badly-seasoned meat which tasted like rabbit, but he hoped was not a human primate. "Tell me how you grew so... old?"

"The same reason why you got so bald, I suppose," the now white-haired, wrinkled-skin, pre-maturely arthritic but still vibrant and defiant behind-the-eyes space traveler said, stroking the top of his now hairless head. "Leave home, get on-the-road disease. Which can be reversed when you...eh." Prometheus held the rest of what he was about to say behind tight lips and a contemplative stare directed behind his ocular portholes.

"Which can be reversed by some miracle medicine Dral can come up with here?" Grim advanced.

"Or we both can develop if you take us home with you to where... the Huntress maybe still lives?" Dral asked, seeing something familiar in Prometheus' eyes. "Who maybe is—"

"—The ideal woman we all envision so we can be disappointed with the real ones that life puts us together with," Prometheus advanced. "Which, if we work those compromise relationships right, can be recreated for us now with our wives, or, if we're good and intelligent fathers, are recreated in our progeny for future generations."

"Right," Dral said, allowing Prometheus to have his lie go unchallenged.

"If you say so," Grim added, seeming to know more truth about the Huntress and Prometheus than the 'gods' ever thought possible. Such was a seemingly impossible realization from a freshly evolved from apes humanoid species on a primitive planet so backward that none of the beings on it knew how to transport themselves to other planets. Prometheus realized there was something about Grim and Dral that he envied now. It was their intelligence of heart. Something that had to be valued. And respected.

"So, it would be interesting if you could tell us a fable from your homeland that had some truth to it," Dral inquired of Prometheus. He helped himself to more 'rabbit' stew to feed his genuinely empty belly, but, thanks to his transformation, he was more hungry to feed his expanding mind.

"And it would be fun to hear one of your fables from your home," Grim added, with a sardonic smile, after which he drank a swig of berry juice that Prometheus had boiled to eliminate the 'firewater' that sedates already lazy-addicted minds. "Yes... Fun. Something that all three of us used to be able to experience. Like... happiness."

"Bliss is better," Prometheus said to his now fellow Comrades. "As we all know, or should."

"Suppose so," Grim said

"Tis what it be," Dral added.

Prometheus rested his weary head on his palm, scratching his beard. He thought about his options. Sharing more common ground with the two 'hybrid' humanoid-aliens he had created would lead him and his Comrades to a Cause bigger than they knew. Perhaps, together, they could fix multiple problems plaguing at least three planets. "You know, it's better to understood than be understood," Prometheus advanced regarding the fellow beings who were lab rats in this noble and necessary experiment. "And better to love than be loved," he put forth, thinking about Athena, who had found her way back home to where she would find someone other than him who would please as well as serve her, so he hoped and lamented. "But... we have to be who we are," he concluded. He pushed down on the soles of his arthritic, then stood up, looking to the sky with fear and longing. "Yes, we have to be who we are. And be where we are supposed to be."

"With our own kind?" Dral asked.

"Or the kind of people we have become?" Grim advanced.

"Which," Prometheus said, possessed by the possibility of the moment. "I could arrange something for... all of us," he proposed. "But—"

"—If you did, you'd have to kill us afterward?" Grim interjected.

"Which would be fine with us, me anyway," Dral said, after which he rubbed the hairless crown of his head. "As long as 'we' can maybe develop some kind of medication that will grow grass over my busy streets. Something I want and, maybe, sort of need."

"Needing and wanting can be the same thing," Grim said. "And should be. Like work and play. If we're doing either one the way we should, then they are the same thing. A merging of the have-tos and want-tos. Like..." he continued, feeling a future time and thought coming to him. "Pleasure and pain being the same thing."

"A quote from Nikos Katzanakis' last novel," Prometheus thought but didn't say. "Nikos being... someone who, well... one of you will be reincarnated into, eventually, after serving so many people who need people. While you are loners, even with someone you love, one of you will write a song that will be both true and popular when you reincarnate into a hippie-redneck music star. You both will plant seeds in children who are ours and ones who were born to other fathers. Human seeds that will germinate," Prometheus thought but...decided not to give voice to. Instead, he remained silent. He pointed to a light in the sky that only he seemed to notice. A beam came down, laying its point on the other side of a hill. He worked his way over to a mound of dirt covering an installation that was supposed to be a launch pad, his deteriorating feet barely able to hold him up. "I've gotta go, and so do you."

"Where?" Dral asked.

"We'll find out when we get there and will meet again," Prometheus said as he hobbled away with hopeful eyes but ever-hurting feet. "Take care of the horses. They're doing all they can to take care of you. And know the way out of this valley, and, well, this—"

"—How will we recognize you?" Grim asked.

"And can we go where you're going?" Dral pressed.

Prometheus stopped, thinking about spilling the beans. Revealing the truth to the two most deserving souls to know it. But as soon as he opened his yapper, set to reveal all of "the gods'" secrets to mortals, he was interrupted by the arrival of a crow. It perched on a thin tree branch, which oscillated to the beat of his pounding heart. "I've been assigned to be somewhere less evolved in ways that matter than this place is, could be, and will be," he said. "But if you can, try and figure out something that will make livers grow fast or make crows prefer to eat anything but that."

With that, Prometheus left his present post, assured that he had done his job. And that his experiment was a success, by the measurements that mattered anyway. But in the war between ignorance vs intellect, cruelty versus kindness, and existence versus Life, he had set in motion a new set of players who, hopefully, would not fuck it up.

ABOUT THE AUTHOR

MJ Politis departed the womb in Hoboken, New Jersey, in 1951. To make good on what everyone who supported, taught and challenged him did, he obtained a Ph.D. in physiology in 1978 which was used to publish 46 research papers in medical journals in reconstructive neurology, toxicology and cancer treatment. He went on to obtain a veterinary degree to extend medical care to fur bearing souls in a wide variety of clinics and cultural settings across the US and Canada. In order to diagnose and cure numerous maladies of the human soul, he obtained an H.B.A.R.P. degree (human being, aspiring Renaissance person) as author of over 80 novels and novellas, as well as producer/director/writer on 27 comedo-dramatic films, which can be accessed through www.longriderpress.net. He has been owned by horses for the last 40 years, currently residing in Interior British Columbia, Canada as home base, regularly commuting to New York to maintain global perspective.

www.ingramcontent.com/pod-product-compliance
Lightning Source LLC
Chambersburg PA
CBHW071424210326
41597CB00020B/3644